CW01655244

THE WHITE DIAL CLOCK

THE WHITE DIAL CLOCK

BRIAN LOOMES

DAVID & CHARLES

NEWTON ABBOT LONDON VANCOUVER

0 7153 6425 1

First published 1974
Second impression 1977

Set in 11pt Garamond, 2pt leaded
and printed in Great Britain
by Redwood Burn Ltd, Trowbridge & Esher
for David & Charles (Publishers) Limited
Brunel House Newton Abbot Devon

Published in Canada by
Douglas David & Charles Limited
1875 Welch Street North Vancouver BC

I dedicate this book to my dear wife for her tolerance and understanding during the many hours of work which its preparation involved

CONTENTS

LIST OF ILLUSTRATIONS

PLATES

FIGURES

TABLES

FOREWORD

This book is about grandfather clocks with painted dials. I use the term 'white dial' as this covers all the various types, although many later examples, far from being white, are extremely colourful. Furthermore, this is the term the dialmakers themselves used at the time. These dials are not true enamel, although they are sometimes incorrectly referred to as such. (Longcase clocks with dials of true enamel are very rare indeed.) The dials have a japanned surface, which is basically decorative paint, hardened by a heat process.

I have often searched for authoritative written comment about this type of clock, and have usually met with disappointment. I found a frequent tendency on the part of southern English writers of the past to pre-date northern clocks, and then to compare them unfavourably with southern clocks of an earlier period.

As I handle clocks of this type every day, a certain pattern of development in styles began to form in my mind. For amusement I made a detailed analysis of 100 clock dials, selected at random, and although I had begun this with some half-formed opinions already, I was amazed at the clarity of the progressive stylistic pattern which emerged: a far stronger pattern than I had expected. Continuation of this analysis gradually threw up a list of features indicative of the period, in other words a set of 'rules', which one could not expect to hold good for every single white dial clock, but which do seem to apply to a surprisingly large proportion of British longcase clock dials. Since British dials were exported to America, it follows that dial-style developments there are almost identical, in the early period at least. So far as I am aware, none of these developments in dial styles are mentioned in any existing horological work.

This is not intended to be a technical book about the mechanics of clocks: there are plenty of books on this aspect already. It is about the ordinary, household, grandfather clock of the type met with in thousands of homes and antique shops.

Some of the views I express may well be controversial, but as far as possible I have tried to support them with factual examples; and a high proportion of the clocks I illustrate are dated positively by the maker. Many of the clocks illustrated are ones which have passed through my hands, and therefore I have been able to make a close personal examination of them. I have tried to use examples which are typical and not exceptional: clocks one might meet with every day rather than rare museum pieces which one might never come across again in a lifetime.

No one person's experience can ever be fully comprehensive, and I do not for one moment regard this work as pronouncing final conclusions. My observations relate essentially to longcase clocks; and while some bracket-clock dials were produced by the same processes and probably the same dialmakers, the developments I outline do not relate to bracket clocks.

If even a fraction of the pleasure I derived from writing this book is shared by the reader, it will have been very worth while.

B. L.

CHAPTER 1

THE YEARS OF NEGLECT

Existing horological reference books tell us little or nothing about white dial clocks, except that their authors usually regard them as very inferior country cousins, almost a different species from brass dial longcase clocks. Some authorities say that the white dial appears shortly after the middle of the eighteenth century; some say 1760; some 1770; some 1780. The disturbing thing about this variance is not so much that it indicates a difference of opinion among experts, as that it is really an indication that no one knows.

In a way this is understandable. Many of the standard clock books were written around the turn of the last century, when white dial clocks were only just out of production. At that time this type of clock, which had been in existence just over 100 years, was not old enough to have an antiquarian interest. In just the same way today we hardly think it important enough to write about furnishing objects made in the early twentieth century. They are not antique, not valuable generally, not in demand. They are simply old-fashioned and in that no-man's land between junk and antique: the sort of thing which our grandparents bought as new for their homes.

In 1894, then, when the first edition of F. J. Britten's *Former Clock and Watchmakers and Their Work* was published ('for more than half a century the standard work on horological history'), brass dial clocks by provincial makers could be bought for minute sums. Few people indeed wanted the white dial ones in this situation. They were dumped in outbuildings, pushed into attics, the cases chopped up for firewood. In the early years of this century second-hand furniture dealers would deliver a white dial clock in perfect order and set it up in one's home for less than £1. In this situation it is not surprising perhaps that Britten did not devote much space to it—maybe half a dozen lines.

Subsequent writers gradually began to include a little information in their books, but the pattern had been set and I cannot recall any book where the author even begins to whet the reader's appetite to own a white dial clock. The precise period at which the white dial clock first appeared was not a problem of great importance then, and does not seem to have been looked at seriously since.

Understandable as this may be, it is none the less surprising when we think a little more about it. One day in the 1760–80 period a man or group of men, or several men independently, had the courage to try to make and sell a clock with a painted dial. It was new, simple, cheap to make, legible; and it became popular. From that moment the brass dial clock died rapidly. The introduction of the white dial was to prove one of the most important and far-reaching innovations in the history of clockmaking in Britain and America. The white dial enjoyed a period of popularity that lasted 100 years. I know that several reference books state that longcase clocks died out by 1820. Don't believe it! They were still being made even fifty years after that date. In fact, by 1820 they were just reaching their height of popularity.

So great was the impact which the new white dial made on the clock trade, that with many provincial clockmakers the change to white dials was not only rapid but permanent. I have data on the clocks of one particular country clockmaker, typical of thousands of similar men, which show that once he began to use the new type of dial (about 1774 in this case) he instantly dropped the use of brass dials, and no brass dial clock by him is known later than the date of his first white dial one. What can be discovered about one clockmaker will no doubt ultimately be discovered about many others, if we take the trouble to investigate.

It seems to me amazing that the introduction of the new white dial has been regarded as a regrettable event by writers about clocks. I can only imagine that this sort of attitude originates from indoctrination through those books on horology which regarded London as the centre of clock design exerting its influence on the rest of the country. True as this may have been in the earlier periods of clock-making, it is no longer true in longcase clock production of the later eighteenth century, when it was in the *provinces* that longcase clocks were really popular. What London clockmakers did in this direction at that time had little bearing on the designs or fashions popular in the provinces. It is perhaps for this same reason that London writers are unsympathetic towards later provincial clock casework, apparently failing to realise that provincial cases of the late eighteenth and early nineteenth centuries fail to resemble earlier London cases, simply because they were never intended to resemble them.

It has been, and apparently still is, fashionable for writers to glorify the brass dial and to decry the painted one because it is not made of brass; though the same people do not normally decry the painted moon dial filling the arch of many a brass dial clock. No one seems to have a kind word for the painted dial. No one, that is, except the countless thousands of clockmakers in Britain and America who began to use painted dials in the last quarter of the eighteenth century to the relatively rapid and ultimately complete exclusion of the brass ones. If the appearance of the white dial did mark the beginning of the end, then that end was almost 100 years in coming, since the painted-dial longcase clock was popular for almost as long as the brass dial one had been. Anyone who looks at a provincial white dial clock of 1800, expecting it to resemble a London brass dial one of fifty years or more earlier, need hardly be surprised at his failure to appreciate it. However, it seems to me to indicate a lack of modesty if such a person ascribes this to an absence of good taste and design in clockmakers and their customers over the best part of a century.

The white dial clock was to find its way into perhaps the majority of homes in the country, cottages and farmhouses as well as mansions and town houses, in a way that the brass dial clock never did. Yet the revolution began so quietly that its originator was soon forgotten, and has remained forgotten. It is unlikely that we shall now establish for certain who made the very first white dial clock, and when. We can, however, make a reasonable assessment of when this took place; I shall cover this aspect in detail later. Occasional attempts were made to produce longcase clocks with dials of true enamel, but because of the problem of applying enamel to such a large surface area, these were not very successful; and when I speak here of producing white dial clocks, I am thinking of the serious and continuous production of them as a regular trade item.

What we are only just beginning to appreciate is that for the vast majority of clockmakers in Britain and America most of the longcase clocks after about 1780 were built with white dials. Brass dials after this date were the exception rather than the rule; perhaps not quite so much applicable to makers in the far south, but holding true particularly of the midland and northern clockmakers. The white dial is far more a sign of the period than the quality, than has yet been suggested in any book.

Quite a number of dealers regard white dial clocks as purely for sale to American, Dutch, German and Scandinavian trade buyers. Some buyers, such as the Americans, buy in vast numbers to try to fill containers and thereby make shipment as economical as possible. Others, like the Dutch and Swedes, come in their own vans on regular trips every three or four weeks and return home with a dozen or so longcase clocks from each trip, along with other antique and semi-antique 'décor' items. Only when one begins to imagine the scale on which this takes place throughout Britain, does one appreciate how many hundreds of clocks leave the country every week, probably for ever.

'Why,' people often ask, 'do so many white dial clocks go abroad?' There are three reasons. First, the grandfather clock was essentially a British product. While some other countries made them with their own national variations, nowhere in the world were they made on the same vast scale as in Britain, where almost every village had its own clockmaker or clockseller; and nowhere in Britain were so many made as in the north of England.

Page 17 *(left)* Thomas Bussell, Cullompton, c 1795; *(above)* William Terry, Bedale, 1787

Page 18 *(above)* Caleb Nicholas, Birmingham, c 1785; *(right)* John Day, Wakefield, c 1780

Second, being more uncommon in these other countries, British clocks fetch a better price there than here, which situation is also helped by the higher cost of living in many of the importing countries. Therefore there is plenty of room for profit, despite costs of transport and duties. Third, and this is perhaps the main reason, white dial clocks (the main type to be exported) simply have not had the same popular appeal here in the present day as the brass dial ones.

Foreign buyers seem to go especially for white dial clocks, partly because they are cheaper than brass dial ones, and because they can probably get a similar price at home for either type. Hence they prefer the type with the higher profit. I have heard it said that the American public tends to prefer painted dials to brass dials anyway. (I am speaking of the average private purchaser, not the collector.)

Some white dial clocks may meet a less fortunate fate, especially if, for instance, the dial is rusty or the worse for wear. Only recently a man asked me if I would get his clock converted for him. It was a thirty-hour brass dial clock, and since he objected to winding it daily, he had decided he would like it made into an eight-day one. I am not a clock repairer myself, but I do know competent men who undertake repairs. I explained to the man that he would be spoiling his clock by doing this, and that if he ever came to sell it, its value would be reduced by his action. But he had no intentions of selling it, he assured me. I explained to him that such a step was not a matter of a minor adjustment, but that it meant replacing the existing thirty-hour movement with a quite different one taken from an eight-day clock (probably a white dial one). He already understood this, but still wanted it done. I gave him the address and away he went. I don't know whether he went to see my clock repairer or not, but the point I am making holds good anyway. Any clock repairer can do the job of converting a thirty-hour clock to an eight-day one. If he does it successfully (that is, cleverly enough to take in the purchaser), there is a good profit margin between the selling price of the thirty-hour clock and the (now) eight-day one. A number of clock repairers who have offered me their services have included the ability to do such conversions high on their list of talents.

It is usually of course the movement of an eight-day *white dial* clock which is plundered for this purpose. And of course when the thirty-hour movement is fitted to the now spare white dial, one still ends up with a thirty-hour white dial clock, of sorts.

Incidentally, this type of conversion is usually quite easy to spot without any technical knowledge. The winding squares of the movement seldom fit plumb into the centres of the newly cut winding holes; and the winding holes themselves are often not truly circular in shape, through careless cutting. If one's suspicions are aroused in this way a quick glance inside the movement will generally reveal a few spare holes in the movement front plate, where the original dial pillars once fitted.

This sort of conversion is not always a modern job. On occasion I have bought clocks which have been in private hands for many years, yet have been through this same process—maybe fifty years ago even. It has often been done to convert one-handed clocks (once thirty-hour ones) to two-handers. I point out this conversion business to illustrate another way in which enormous inroads have been and still are being made into the numbers of white dial clocks.

The nearest similar situation of which I have had personal experience, has been where a customer has offered to buy clock A if I would switch it into case B. This happens every now and then, but so far I have always resisted. On another occasion, a customer who wanted to buy a white dial clock but did not like its matching steel hands, asked if I would replace them with matching brass ones. This I refused to do. 'Oh, but I'll pay you for putting the other hands on,' he insisted. I explained that if he wanted to spoil the clock once it was his, I could not stop him; but I certainly had no intention of helping him to spoil it.

In recent years the white dial clock has suffered in yet another way. Probably the biggest single way in which it has suffered is in having had its cases stolen, in the same way as its eight-day movements, for the benefit of the brass dial clock. It is simply not correct to imagine that in its day the white dial clock was regarded by the clocksellers or clockbuyers as in any way inferior to the contemporary brass dial one. The white dial clocks were the newest and most fashionable kind. This contemporary view of its equality, if not perhaps its superiority, is supported by its having been made to accompany some really splendid cases. Over the years very many of these fine cases have been taken to rehouse brass dial clocks to satisfy the modern customer's desires for fashion. It happened a great deal in the past and it still happens frequently today.

Eventually a good white dial clock in a good original case will be a vastly rarer thing to obtain than a similar brass dial one. Then perhaps the situation will begin to reverse itself, and cases will be stolen from brass dial clocks to be put on to white dial ones. Not that I would advocate that practice either.

These then have been the fates of many white dial clocks over the years, and some of these practices still happen. There are signs now that the tide is turning, and very rapidly, as antique clocks become scarcer almost by the hour. People today are much more conscious of their responsibilities towards the past, and there is a far deeper and more widespread interest today than at any time before, in antique objects of all kinds. Perhaps the impersonal nature of modern mass-produced goods causes us to find greater satisfaction in and affection for the works of earlier times, where we share something of the pride and pleasure which a craftsman took in his work. White dial clocks are now beginning to take their rightful place in the scheme of things; and while few will appeal to the old school of 'collectors', vast numbers will find appreciative homes among this wider public which loves antiques for their own sake, and, most of all, which *cares*.

CHAPTER 2

THE BEGINNINGS OF THE WHITE DIAL

Before the introduction of the white dial, all longcase clocks in this country had dials made of brass, usually with separate features such as chapter ring, name boss, etc, superimposed on the dial sheet. The whole dial was made of brass; but in general any engraved part was silvered over when new. The effect was that the dark, 'blued' steel hands (I have never seen original *brass* hands on a brass dial clock, by the way) stood out well against the silvered background, on which engraved details such as hours and minutes were filled with black wax for clarity. This was the form of the longcase clock from its very beginnings.

Later in its development, however, a new type appears. On this the dial is made out of a single sheet of brass with all the details engraved on it, and the whole sheet silvered over—a one-piece silvered dial. This appears always to have been regarded as a natural development of the ordinary brass dial clock, arriving on the scene late in the eighteenth century, though not many authorities seem anxious to suggest a date for its appearance. It does appear round about the same time as the white (painted) dial clock, and at a distance the two resemble each other superficially.

The one-piece silvered dial seems to be largely a product of the south of England. I have seen very few north-country ones of this type and period. Here I am thinking of normal longcase clocks, not austere regulators and regulator-types of clock, which are usually timepieces (ie non-striking) only, and which often have circular dials. It should be explained that a regulator was a clock made particularly for its precision in timekeeping, often with complicated escapements for greater accuracy or special devices to counteract the influences from temperature changes. Regulators were sometimes made by the clockmaker for his personal use in setting the timekeeping of his other clocks. Sometimes they were made for offices and public buildings, for scientists and astronomers, or wherever extreme accuracy was especially demanded. By their very nature they were uncommon items, and while white dial regulators do exist, not only are they seldom met with, they are far from typical of clockmaking of the period. For this reason I have not illustrated one in this book, where my aim is to show what is normal and typical.

I show as an example of the one-piece silvered dial, a clock by Thomas Bussell of Cullompton, Devon, made about 1790–1800. It has steel hands of matching pattern and apparently original to the clock: hands of this type would be unusual before 1790. However, some one-piece silvered dials have the earlier type of non-matching steel hands, as on the earliest white dial clocks.

As far as I am aware, those books which mention the subject at all have suggested that the painted dial began its life as an imitation in cheaper form of the one-piece silvered dials of the Thomas Bussell kind. This is an understandable view. Longcase clocks had always had dials of brass, and it is perhaps reasonable to assume that such a clock with a dial in a new material succeeded those in the traditional material. This view is logical but it has not been proved; and it is one with which I certainly disagree.

Such a view of course fails to take note of the fact that the one-piece dial was almost exclusively a product of the south of England. I am not suggesting that the one-piece silvered dial was unknown in the north; but it was very uncommon there. In this situation how could northern clockmakers readily take to the painted dial as a cheaper *alternative* to a type of clock virtually unknown to them? That the northern clockmakers did take to the painted dial, there is absolutely no doubt. What I would question is whether they took to it as an *alternative* form.

As we shall see, the white dial was already being marketed by 1772. The question I am raising is whether the one-piece silvered dial was known as early as that. I personally do not know of examples so early, and certainly most of the one-piece brass dials one sees are of the 1790–1800 period or later. It is possible that *after* the introduction (about 1770–2) of the painted dial, there began an immediate attempt on the part of dial engravers to copy this new white dial style but in the traditional material (brass) and thereby to retain the engraving work which they had always been accustomed to receive from the entire clockmaking industry. Faced with the possible loss of such a large portion of their business, the engravers could naturally be expected to retaliate by attempting to make dials which looked like the painted ones but which had the durability of brass.

As to which type came first we cannot be sure, unless a one-piece silvered brass dial comes to light that can be proved to be a pre-1770 one. Neither is it very important which type came first, except in relation to the stigma which has always been attached to the painted dial clock: that of its being an 'inferior imitation' of a brass dial one. It is this accusation (without trial or proof) which has led to an almost total lack of interest in white dial clocks, on the part of informed clock enthusiasts, for almost 200 years.

If, as a parallel example, I could prove that the one-piece silvered dial was in fact an imitation of a painted one, that would be no argument for rejecting one-piece silvered dial clocks as being unworthy of attention. It would seem to me to be much more sensible to treat each type independently of the other, for certainly the dial of both types is simply covering up examples of the *same* kind of clockwork. At the time we are concerned with (1770–90), I do not think anyone would suggest that the movements of ordinary longcase clocks would be anything but similar, regardless of which form of dial they were attached to.

The cases of longcase clocks from, say, 1770 onwards, were no better on brass dial clocks than on white dial ones. If, therefore, both brass dial and white dial longcase clocks from 1770 onwards have movements which are largely indistinguishable one from another, and if they differ only in so far as the dial is concerned, then surely to class either one as superior to the other is to judge in a superficial way.

I find it very difficult to accept that the beauty of engraving surpasses the beauty of painting. If this were so, would not every home be filled with engravings rather than paintings? In fact, the popularity of paintings greatly surpasses that of engravings and always has done, and this very factor may have been one reason why the painted dial clock rapidly replaced the brass dial one.

As I see it, the white dial originated as a new type of clock dial which not only served its purpose in the same way that the traditional brass dial had done, but in fact did the job better. So well did this new dial perform its function, that it rapidly outgrew the brass dial in popularity to the extent of the eclipse of the previous kind. Surely if it had done its job less satisfactorily than the brass dial, it would never have caught on at all.

Before outlining the known facts I must make reference to an example of a white dial clock supposedly made in 1761. In his book *Chats on Old Clocks*, published in 1917, Arthur Hayden illustrates a white dial clock 'dated' 1761. It has a most unusual dial of oval shape. The only other similar one I have ever seen is that by James Lomax of Blackburn which I illustrate later in this book. Of course, a dial of this type could have been made in any shape at all. Perhaps tradition and commonsense established that the white dial was most often made in square and arched shape, and somewhat less frequently in circular shape. I say commonsense, since a square and even arched dial is far simpler for the casemaker to deal with than is an oval, egg-shaped one of the type which Hayden shows, or even a circular one. The oval dial illustrated by Hayden bears the name of Marston of Salop. The date of 1761 is claimed for it by virtue of a brass plaque on the case door bearing this date, but it should be borne in mind that a clock dated genuinely by a plaque attached to it is very rare, and evidence of this kind should be regarded as doubtful, in the absence of other proof of the clock's age.

The clock by Marston of Salop is typical in its case and dial features (except for the dial shape) of about 1780–5. The only known information on the maker is that he was listed in the directory of 1795. Those authorities who give his date as 1761 do so on the basis of the clock shown by Hayden. I would put the date no earlier than about 1780, and certainly it is *not* proof of a white dial clock's having been made as early as 1761.

Some longcase clocks exist which have dials of true enamel, but they are exceedingly rare. This real enamel is a white opaque glass fused on to thin copper at red heat. True enamel dials had long ago appeared in watches and, I am told, by 1750 they were standard. On longcase clock dials, however, enamel was not a success because of the very large dial areas involved. Both sides of the dial had to be coated to prevent distortion, and some panelling to give strength was desirable, though this was not possible for longcase dials.

These rare examples of true enamel dials seem to be in the period 1770–80. One exists by John Green of London, having the arched portion made separately and then attached to the square, presumably to reduce the size of the individual sheets for enamelling. Another interesting example is thought to have been made by Richard Comber of Lewes: the name has worn off over the years. Made about 1770, it has a three-train movement chiming the quarter hours on twelve bells and striking the hours on a thirteenth bell. It is basically a brass dial clock with corner spandrels, but it has a circular, enamelled centrepiece for the numerals and a smaller, similar one in the arch for a strike/silent feature. In other words it reflects the beginnings of the move away from the brass dial to the more legible white form. This clock is an attempt at combining the best features of both the brass dial and the white dial. From the photographs I have seen of it, it is a superb clock in a similar case far above average and must have been made for a wealthy customer. The *painted* white dial clock on the other hand had enormous potential because it was so very cheap to produce.

Recently I saw a clock combining the two kinds of dial in a different way. This was an eight-day, arched dial clock by Dollif Rollison of Halton, near Leeds, Yorkshire. The dial sheet, painted in a Chinese design, was deep red with gold decoration in the manner of a red lacquered cabinet. Superimposed on this dial were a *brass* chapter ring, a brass date ring and a brass seconds ring. The appearance was most effective; and this seems to be a different approach to the same problem of combining the best features of both brass and white dials. Unfortunately, no dialmaker's name was marked. The clock was probably made about 1775. Such a combination of dial types is very unusual.

All we can deduce is that while attempts were made at true enamel, such dials were so unsatisfactory or perhaps so costly to produce in commercial numbers, that the idea was soon dropped. Many owners of white dial clocks describe the dials as being 'enamel', 'porcelain', 'Devonport porcelain', 'Battersea enamel', and so on. The vast majority of these are mistaken descriptions by the owners, perhaps not surprisingly so. In 1799 the famous cabinetmaking concern, Gillow's of Lancaster, ordered a thirty-hour clock dial and movement from Newby's of Kendal, at a cost of £2 11s 0d, to go into a case they were making; and they describe it as a 'china dial'. Newby's would almost certainly have obtained this dial from the Birmingham manufacturers, as I shall explain later. These mistaken identifications simply show how realistic these painted dials were 'in imitation of enamel'. Perhaps such descriptions as 'china dial' were deliberately inaccurate sales names used to make this new cheaper dial seem to be a more delicate and more choice item than the term 'painted' would suggest. Longcase clocks with true enamel dials are so rare that the known examples can probably be counted on the fingers of one hand.

The emphasis in this book is upon white dial clocks, so let us imagine that we are looking at clocks of this kind for the first time. Let us further imagine that we have never read anything about them before.

The story begins in Birmingham in 1772. In that year the *Birmingham Gazette* carried the following advertisement in its issue of 28 September:

White Clock Dials
Osborne and Wilson, manufacturers
of White Clock Dials in Imitation of Enamel, in a Manner entirely new, have opened a Warehouse at No. 3, in Colmore-Row, Birmingham, where they have an Assortment of the above-mentioned Goods. Those who favour them with their Orders may depend upon their being executed with the utmost Punctuality and Expedition.
N.B. The Dial Feet will be rivetted in the Dials, and such Methods used as will enable the Clock Makers to fix them to the Movements.

This is the earliest indisputable evidence I know of regarding the making of white dials. It is reasonable to assume that white dials cannot have been made for long before this advertisement appeared. If they had, Osborne would hardly have been able to describe them as being made 'in a Manner entirely new'. We can assume, then, that white dial clocks cannot have been made much earlier than 1770, even 1772. Further evidence in support of this argument is that no white dial clock is yet known which can be proved to have been made before 1772.

The last line of the advertisement is an interesting one, in that an explanation is given that the dial feet will be riveted into the dials, yet in such a way that they can be attached to the movements without difficulty. Of course the dial feet *had* to be fixed to the dials *before* they were 'enamelled', so that the ends of them were hidden by being covered with the paint finish. On a brass dial clock the feet could be fixed by the clockmaker anywhere into the dial where they best fitted in with the movement frontplate. Very often the ends of the feet met the brass dial behind the chapter ring and so were hidden.

The difficulty with ready-fixed dial feet was that they might fit in with the movement at an awkward spot, where they were perhaps in the way. To set their customers' minds at rest on this point, Osborne

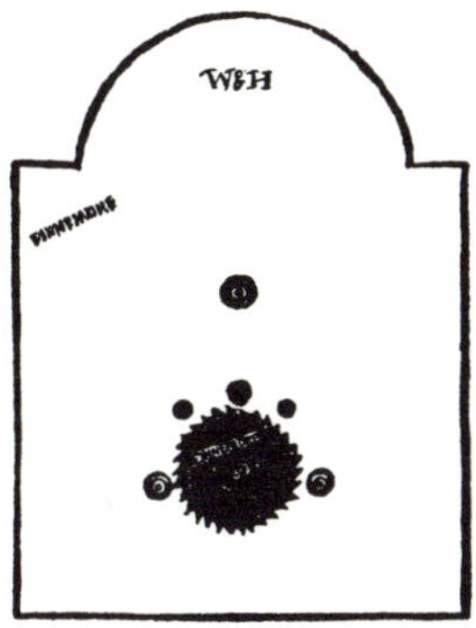

1 (*top left*) A typical falseplate, showing three feet for attachment to movement frontplate, holes to take feet from dial, and cut-away sections to allow for clearance of moving parts of the clock. Also shows name of dialmaker

2 (*top right*) Exploded diagram showing normal fitting of dial *without* falseplate to movement, in this case by three long dial feet

3 (*bottom left*) A falseplate dial (exploded) showing short (four) dial feet attaching to falseplate, which in turn attaches to movement frontplate by three feet

4 (*bottom right*) A dial seen from the back, showing positions of normal name-stampings either in arch or in corner and also showing a date disc with name-stamping: one would not normally see the name in all three positions on the same dial

& Wilson specify that there will be no problem in this respect. They do not say how, but it must have been done by using a 'falseplate', which is an intermediate plate-fixing between dial and movement. The falseplate is explained more fully later in this book, but it is interesting to note that by no means all the dialmakers who followed on after Osborne & Wilson used falseplates, at least not all the time, as one sees considerable numbers of dials without falseplates for long after 1770. Perhaps it was that dials without falseplates (ie with long standard dial feet fitting directly to the movement) were supplied by those dialmakers who also supplied the movement to go with them.

Five years later, Osborne and Wilson decided to part company. In the *Birmingham Gazette* for 29 September 1777 they announce that the partnership is being dissolved and that Wilson will carry on alone:

> Birmingham Sept. 29. 1777.
>
> The Partnership lately subsisting between Messrs. T. Osborne and J. Wilson, Manufacturers of Clock Dial Plates, was on the 27th Instant, dissolved; of which the Public are requested to take Notice; and also that this Branch of Business will be carried on by Wilson alone, at No. 11, Great-Charles-Street, Birmingham, who hopes for the Continuance of the Favours of his Friends and the Public . . . All Persons having any Demands on the said Partnership, are desired to send an Account of the same, as above; and all Persons who stand indebted to the aforesaid Partnership, are desired to pay the same.

It was maybe that Osborne was failing in health. However, in the issue of 19 January 1778, we learn that Osborne has decided to carry on after all:

> *Clock Dials*
>
> The Partnership between Thomas Hadley Osborne and James Wilson, Manufacturers of Clock Dials, is this Day mutually dissolved:—They have authorised Mr. John Walker,

of Colmore-Row, to receive and discharge all Debts due to, and by, the said Partnership. Birmingham Dec. 24, 1777.

T. HADLEY OSBORNE, thinks it necessary to aquaint the Public, that he intends to carry on the Manufactory of CLOCK DIALS as usual, when the Favours of his Friends will be punctually attended to, and gratefully acknowledged by their Obedient humble Servant,

T. Hadley Osborne.

No. 20, Cherry Orchard, Birmingham.

JAMES WILSON respectfully informs his Friends and the Public, That he continues the Clock Dial Manufactory, (late Osborne and Wilson's) at No. 11, Great-Charles-Street, where Orders will be gratefully received, and executed in the most expeditious and complete Manner, by their obedient Servant, Birminghar , Jan. 5, 1778. JAMES WILSON

Evidently this indicates a change of heart on the part of Osborne, since the earlier notice said that Wilson would carry on *alone*. It would seem that by now a once amicable partnership was turning sour; and henceforth these men must have been determined rivals. Osborne then continued his own business, but not for long, as he seems to have died in 1779, after which time his widow, Ann, continued for some years.

James Wilson, however, appears to have prospered until his death in 1809. Recently I was able to track down his will, dated 20 March 1809 and proved at London 15 June 1810. In it he describes himself as a clock *dialmaker* (not a clockmaker), and this appears to be confirmation of the fact that he supplied dials only, not complete clocks. He seems to have been a widower as he mentions no wife. His two daughters and one son were minors in 1809. Among those appointed trustees and executors was one Edward Simpson, japanner, a man who produced applied painted decoration of the type used on a white dial clock. Osborne and Wilson,* then, probably without realising it, started a revolution in the clock world. We cannot say for certain that they were the very first white-dial makers, but they are the first we know about. Instead of becoming famous their names were forgotten the year they died, and it is doubtful whether even one clock enthusiast in a hundred has ever heard of them.

* Since I began writing this book, my attention has been drawn by Mr Kenneth Roberts of Bristol, Connecticut—himself author of several works on American horology—to an article in *Antiques* in September 1931 by Penrose R. Hoopes, entitled 'Osborne & Wilson, Dialmakers'. The article is written with special reference to the use of these dials on American clocks. An article in *Apollo*, vol 48 (1948) by W. A. Seaby also refers to them briefly. I mention these two articles to acknowledge that I am not the first to write about Osborne & Wilson. I believe that Hoopes was mistaken in his assumption that falseplates were introduced to prevent chipping on the dial feet ends, as I shall explain more fully in Chapter 3.

CHAPTER 3

THE DIALMAKERS

Most people are aware that the clockmaker did not make the case: he bought it ready made from a cabinetmaker. What is not quite so well known is that he did not usually make the dial either. Dial-making was a specialist business and it was seldom the clockmaker's own province. This applies to brass clock dials, where the engraving was done by a specialist engraver, possibly working to the clock-maker's instructions. Engraving was a highly skilled trade, and in relatively few instances would a clockmaker also have the skill to engrave his own dials.

Now when the white dial clock appears about 1770, from its very beginning dial supplying is a separate trade. Because dialpainting is so obviously quite distinct from clockmaking, many people are already aware that such dials are 'factory' made, and that they simply had the name of the retailer painted on them. But they often regard this as an indication of the inferiority of the product, being quite unaware that brass dials could also be said to be 'factory' made, in just the same way that pinion wire, spandrels, dialplates, hands, and even complete clock movements could all be bought in by the retailer from the trade suppliers certainly by the 1770s and for a good number of years before that too. By a 'factory' is meant a specialist workshop, the size being not so important as the fact that they were specialist component suppliers.

When they first made the new white dial, the men who had the ingenuity to produce it also hit upon another idea to help sell more of their products. This was to sell the dials in such a way that almost any normal movement would fit them, regardless of who had made or supplied the movements. In other words, the smaller clockmakers in country areas, the men who worked alone or who, by traditional methods, made much of their own clockwork, could now if they wished buy these new white dials in just the same way as could the larger town clockmakers who may have been using standard-sized movements (that is, 'factory'-made movements). The smaller countryside makers were of course the men in whom true 'craftsmanship' lingered by virtue of their making much of their own product; while the larger town makers, who employed several apprentices and journeymen, merely kept a supervisory eye on their workers, and might never have laid a finger on the product which bore their names. This universal fitting was achieved by use of an iron 'falseplate'.

The traditional dial (white or brass) fitted on to the frontplate of the movement by means of three or four feet of about 1in in length. The disadvantage of these of course is that in trying to fit any movement to such a dial which was not purely designed for it, one finds the feet cannot always fit into the frontplate because the movement pillars (or some other part of the movement itself) are in the way. By attaching a falseplate between dial and frontplate this problem is overcome. The falseplate is already fixed to the dial when supplied, naturally on shorter (½in or so) dial feet. Then by drilling holes at any convenient positions into the falseplate one can attach almost any conventional movement without problems.

The falseplate system of fitting the dials has another advantage in that there is less tendency for the short-fitting feet to bend and thereby chip off enamel at the point where they are attached— something which not infrequently happened with the longer fitting feet of dials without falseplates. This advantage of the shorter dial feet preventing chipping was only a secondary one. It was not, as is sometimes stated, the main reason for the use of falseplates. Osborne and Wilson make this clear in their advertisement of 1772: that the main purpose of the falseplate is for ease of attaching the move-

Page 35 (*left*) John Bancroft, Scarborough, c 1785; (*above*) James Lomax, Blackburn, c 1790

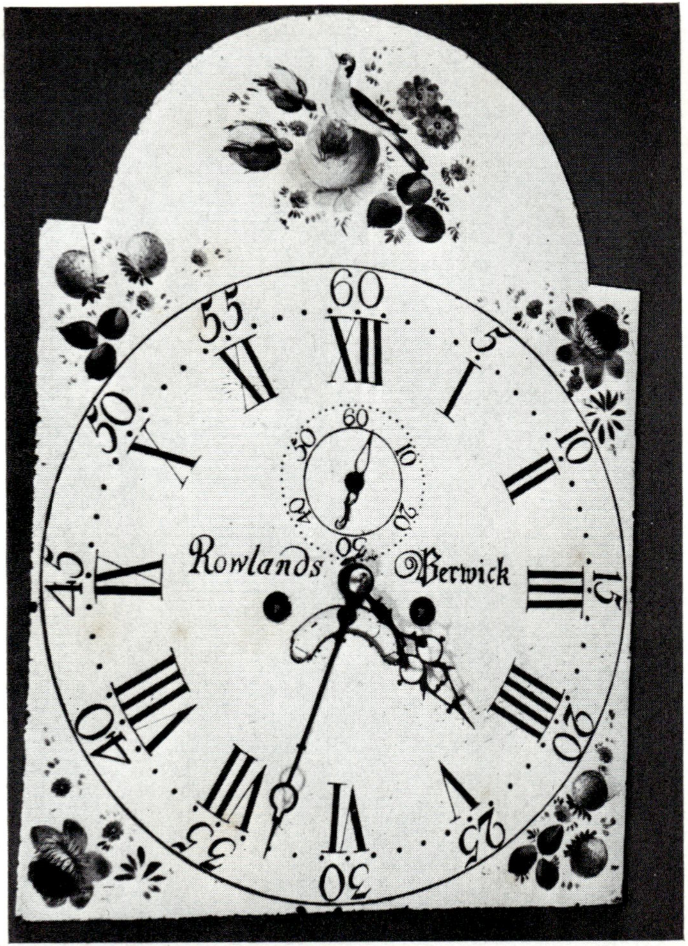

Page 36 (*left*) Rowlands, Berwick, c 1785; (*above*) Forster, Sheerness, c 1790

ment. Further evidence that the primary purpose of the falseplate was not for the prevention of chipping, can be seen from the fact that falseplates appear only on eight-day clocks and never (or exceedingly rarely) on thirty-hour ones of the same period—this despite the fact that chipping is far more likely to occur on the normal three feet of a thirty-hour dial as opposed to the normal four feet of an eight-day one. The falseplate was attached to the dial to accept the movement more easily, not, as is sometimes incorrectly stated, attached to the movement to accept any dial.

The benefit as far as we are concerned in studying the age of clocks rather than the technicalities of them is that the falseplates very often have the name of the dialmakers cast into them. Quite a lot can be discovered about some of these dialmakers, most of whom worked in Birmingham. The falseplates can often therefore be of assistance in dating a clock.

One point I had best clarify before going further. It is not true, as is sometimes stated, that the falseplate dial will fit *any* movement. This depends ultimately on the position of the winding squares and hands in relation to each other, since the four points involved must relate to each other at the same angles and positions in the movement as the four holes in the dial (hole for seconds hand, hole for main clock hands, two holes for the two winding squares). This applies only on an eight-day clock, as on a thirty-hour clock there is only one hole, that for the main hands, there being no winding holes on a thirty-hour clock and very rarely a seconds hand. On an eight-day dial, then, four holes must line up with four points on the movement; on a thirty-hour dial only one hole must line up. In practice a falseplate dial will fit many types of movement, while the normal dial without a falseplate will fit only the movement which was made for it (or one very similar to it).

Birmingham was the main centre for the manufacture of white dials. The dial is basically a sheet of iron of the required shape: this generally followed the traditional shape of brass dials—ie square or square with an arched top, which I refer to as an arched dial, though sometimes they are called 'break-arch dials' or, less appropriately, 'broken-arch dials'. 'Arched dial' was the term used in the eighteenth century by the dialmakers themselves, which would seem to be another good reason for us to keep it. Unusual shapes are sometimes found. One occasionally sees circular dials, which were popular in south Yorkshire, Derbyshire, Scotland and perhaps in other localised areas too. A very unusual egg-shaped oval dial is illustrated in Arthur Hayden's *Chats on Old Clocks*, one of only two that I know of with this strange shape. The other is illustrated in this book (see p 35). The dial of course could have been *any* shape at all, but sensibly, square and arched dials account for the vast majority.

The iron sheet was treated on the face side with a red-lead type of base paint, presumably to prevent rust. The back was often painted black. One can sometimes see the runs of red-lead paint spilling over on top of the black backing. Then the dial face was painted white, probably by being dipped into a vat of prepared paint. It was not enamel in the true sense, as on the dial of a watch. It was a paint of a hard and durable nature baked on by a heat process. Whatever state of deterioration these dials may now be in, one very rarely seems to see the white background *worn* off. It sometimes chips off, it sometimes shells off through rust, but it never seems to wear off with normal use.

The painted scenes superimposed on this white background, and the painted numerals, were obviously of a far less durable finish. One does see these wearing off, the numbers especially. It is confusing to call such a dial an enamel dial, and similarly confusing to call it a painted dial, though both terms are frequently used to mean just the same thing as a white dial. The makers themselves referred to it as a white dial, and this is probably a further indication (as we shall see later) that the earliest dials were of the all-white or largely white kind.

Before I list the names of known dialmakers, I must explain one or two points about the sources of this information and the dates, since this is vital to an understanding of how to use the list. Most of the dates for these men come from directories, and because of this they do not indicate either the beginning or the end of a man's career. The fact that Christopher Wright is known to have been working from 1835 to 1842 does not mean that he could not have worked both before 1835 and after 1842. It normally means that he is listed for the first time in the 1835 directory and not in the previous one of 1830. He could possibly have been there by 1831, or even before then, if by any chance he had been omitted from the earlier directory.

Quite possibly people like Wright of Birmingham were subject to erroneous entries in directories, and without the convenience of twentieth-century communications it was probably difficult to have such errors corrected. I know of instances of clockmakers still being listed in directories two or three years after they had died. It is therefore essential always to appreciate that lists of this type are only a guide and that the dates given are not definite.

Sometimes the falseplates themselves have no names on them and in these cases the dialmaker's name is often found stamped into the back of the dial itself, usually in the top left-hand corner, looking from the back. Now at first glance many dials show no signs of having a name stamped into them on the reverse. But very often the reverse is covered with a layer of thick black paint, so thick that it can easily obscure such a stamping, and sometimes the stamped name can just be made out faintly beneath the paint and rust.

On dials with an aperture date (ie a painted date disc showing through an aperture) the back of the date disc is very often stamped too. The anonymous dial of a Dawlish (Devon) clockmaker (see p 82) is stamped this way on both the dial and the date disc—made by Wright's of Birmingham. Some dials have unstamped falseplates, unstamped datewheels and no stamping on the dial itself. Perhaps some dialmakers did not care to leave their marks.

On the subject of unstamped dials another possible explanation occurs to me, but one which is rather difficult to investigate. When a dial was being made, prior to painting, the front and the back were interchangeable. The side eventually to be the dial face was coated at least twice with various paint layers prior to the application of coloured paint. On those dials which are name-stamped, the name is usually in a corner or in the arch. If the white background were applied to the wrong side (ie the side with the name-stamping on) this name would be obscured in the process and the slight impression remaining would be hidden by the final hand-painted details. In other words, many of the apparently unstamped dials may in fact have names *beneath* the dial surface. It is a theory which is difficult to test, as one cannot scrape down dial faces to find the answer. But at least it is a theory worth considering. It seems to me more of a possibility than that a dialmaking firm did not bother to mark its products with identification (and therefore advertising) marks of some sort.

By no means all dials had falseplates. They seem to be more frequently found on eight-day clocks; very rarely on thirty-hour clocks. In fact, I know of no example of a falseplate on a thirty-hour clock. As I have already said, the latter were less of a problem, having only a single hole in the dial. If we assume that movements were built to suit the pre-purchased dial rather than dials being bought to fit to pre-made movements, then it was an easier task fitting one to the other, whether of the eight-day or thirty-hour type. It would seem to me from records I have examined concerning the stocks of two northern clockmakers that this was the system operated, a stock of dials being held and movements made up to fit the dial of a customer's choice.

It seems possible that most early eight-day dials were sold with falseplates, to encourage the smaller clocksellers, who still made their own movements, to use them. Later on, particularly after about 1830, one seems to see fewer falseplates on dials. Maybe the later ones were sold as complete clocks—dials and movements together—in which case, of course, the need for a universal fitting was removed. I must not make too much of this point, as it is one where further research is needed before a positive solution can be established; but it can be said with certainty that many early eight-day clocks have falseplates and many late ones do not.

Birmingham was not the only centre where dials were made, but it was the main centre. At one time I found it very hard to accept that local clocksellers in remote parts of the Yorkshire Dales, for example, would even dream of sending halfway across the country for their dials when there were dialmakers nearer to home. I know from studying many white dial clocks that they did buy Birmingham dials. As we shall see in the chapter on white dial clocks in America (Chapter 12), the Birmingham dialmakers could draw in orders from much farther afield than Yorkshire, and they shipped their dials not only halfway across Britain but halfway across the world. This fact must be a measure of the successful selling techniques of such concerns as the Osbornes, the Finnemores and the Wilsons whose false-plates one sees frequently. How many engravers of brass clock dials could claim to have sent regular consignments of engraved dials across the Atlantic? Was there even one?

It must also be largely because of this centralised industry that the orderly progression in dialpainting styles took place. If local artists here and there throughout the length and breadth of the country were all busy painting dials, there could be no steady development in dial styles, but an absolute hotchpotch of them expressing the whims of each individual artist.

We cannot be certain that all dials were sold in a fully finished state. They could have been sold numbered and not painted. They could have been sold as blank white sheets. It is likely, however, that the majority were sold fully finished, as this is consistent with the orderly progression of painting styles, which I shall examine in detail later. Recently I discovered a dialmaker's order form, pasted on the back of a dial of c 1785 by Hepton of Northallerton, Yorkshire; and this suggests that the type of painted decoration was specified in the order from the clockmaker to the dialmaker, thereby supporting the idea that the dials were sold fully finished.

At this time retail clockmakers or clocksellers would not generally make up a large selection of clocks to sell from a showroom. Rather they might have the odd one or two ready, and on sale for a chance customer, but normally they would make a clock, or buy one in, for a specific order. They must have had some method of describing to a customer the range of available dials. If this was by means of printed leaflets showing dial patterns, none seem to have survived.

Perhaps an exception to my comment about clockmakers not

keeping a large stock is in those instances of men to whom clockmaking was a secondary business. There were very many of these. Some clockmakers were shopkeepers having such goods as ironmongery, hardware, umbrellas, barometers, jewellery, silverware, watches of course, rings, cutlery, buckles and fancy goods of this kind. This type of merchandise has long been associated with the clockseller's business; and I do not mean to classify *these* men as part-time clockmakers, although clocks formed only a part of their total trade. Recently I saw a printed label pasted on the inside of a clock door sold about 1830 by Richard Blakeborough of Otley, Yorkshire. It gives an indication of the wide variety of goods handled by a man who appears in the directory as a 'clockmaker'. The list includes: hardware, jewellery, weatherglasses, bottlejacks, ear-rings, thimbles, cutlery, spoons, guns, pistols, pans, kettles, files, saws, hammers and other tools, brushes, copper scales and beams, teatrays both paper (ie papier mâché) and japanned, japanned bread baskets, umbrellas, knobs, escutcheons, locks, keys, hinges, 'stringing & shells for enlaying' (ie for cabinetwork), walking sticks and canes, fifes, flutes, violin strings, looking-glasses, black lead pencils, rat and mouse-traps, scythes and sickles, and . . . 'the best price given for old gold & silver'. Many of these items would normally be associated with the stock in a nineteenth-century clockmaker's shop, though such a dealer might seem to us today more appropriately described as a hardware dealer. Richard Blakeborough, however, was not a part-time clockmaker. John Manby of Skipton, Yorkshire (1815–50) sold an equally varied range of goods.

By a part-timer I mean a man whose main occupation left him certain quiet periods, during which he made clocks. Examples I can think of are the farmer-clockmakers, of whom there were many. During the quieter winter months they worked on clocks and perhaps built up a stock which they would take to sell in the markets when the finer spring weather arrived. Men like Will Snow of Padside, near Otley, Yorkshire and his three clockmaking sons were first and foremost farmers, and clockmakers in their spare time. Some were blacksmiths, seaside boarding-house keepers, or cabinetmakers. Men in all kinds of trades made clocks to provide a supplementary income. This is why so many clocksellers appear in tiny out-of-the-way villages where you might think they would not sell one clock in a year; and the clocks they made need not be in any way inferior to those made by a full-time city clockseller and his employees. This type of farmer-clockmaker would make up stocks ahead of his requirements according to the time he had available.

The fact that many of the examples I illustrate are of Yorkshire clocks is because I have tried to select examples of each particular style, and in some cases could not obtain illustrations from each part of the country of exactly the datable types that I wanted. However, what is true of one area in dial styles is equally true of another, with the proviso that Yorkshire was the largest clockselling county of all, and that white dial clocks were generally more popular in the northern, than in the southern, counties of England. However, because of the main concentration of dialmaking in Birmingham, the styles of dial are not so very different whatever the name of the 'maker' and the town written on them. The chapter on American clocks (Chapter 12) will illustrate this point even further.

But I am not suggesting that dials are the same all over the country at any one time. Southern clocks tend to have smaller dials than northern ones, and their styles progress more slowly in some respects. East Anglia, for instance, seems to have had a liking for the almost-all-white dial, like that by Juler of North Walsham, Norfolk (see p 81). Such dials occur in Essex too, but would be very unusual on, say, a Yorkshire clock.

What this means of course is that if there were fashions or preferences in certain parts of the country, then the dialmakers must have been wide awake to this fact, and must have known which type of dial would sell in which areas. Alternatively we must assume that each clockmaker specified the dial style each time he placed an order.

Another feature is that one never sees the same dial twice, at least I have not done so personally. One sees dials which resemble each other in various ways, sometimes closely, but with some variation in the painted decoration. How much easier it would have been for a 'factory' system to have had people sitting in a row, each one repeating the same design over and over again. They did this in the pottery factories; but not with clocks. Was this done intentionally to make each clock unique? Presumably it was.

In the pottery factories, of course, the designs on many items were transferred on, sometimes with additional hand-painting over the transferred pattern. This is not usually the case with clock dials, though there are exceptions. Very occasionally I have seen a dial with transfer patterns. One I can think of depicted hunting hounds in the corners and a hunting scene in the arch, apparently very casually touched up here and there with hand-colouring, but the transfer had been left plain for the most part. The result was very effective and it is surprising that more dials were not decorated by this method, which must have been simpler and cheaper than hand-painting. Perhaps it was avoided because of this very factor of repetition, and because white dials were intended to be 'one-off' jobs, each one being unique.

While still on the subject of transfers, the following item occurs to me. Surrounding the painted areas on most clocks one can usually see what has been described to me very effectively as a kind of 'halo'. One sees it around a painted corner decoration and also around or sometimes just below the arch scene. It shows up mostly where the painting ends and the white of the dial begins. This halo, which is present on very many painted dials, might seem to suggest that the painted scene is transferred on, as it resembles the sort of halo around a transfer. All it is in fact is the edge of the coat of varnish which was applied over the painted scene to seal it. The varnish discolours at a different rate from that of the white area of the dial, and after the years it becomes a different shade, usually yellower. The varnish was applied originally over the painted areas only, not over the entire dial. This is why the numbering and lettering frequently wear off while the paintings are less prone to erasure.

Now some readers, I hope, may challenge my statement that one never sees the same dial twice. Perhaps I had better enlarge on this a little. In the early period (Period One), with a square dial clock particularly, I accept that dials can look very similar to one another. There are often only the painted corner decorations by which to judge. The numbering certainly will be similar, and the corner decoration can often be a single flower, such as a pink rose or a spray of flowers. In the 'gold spandrel' type of dial, as in the Terry and Nicholas clocks illustrated (pp 17, 18), there is naturally not a great deal of scope for variety. I suppose, therefore, that it is by no means impossible to find two identical dials belonging to Period One.

I was really thinking of the post-1800 arched dial clocks which have painted scenes in the corners and arch. The same themes certainly recur, but in many varied forms and styles, and even two dials showing the same theme will show it with some variations. I said that I hoped some reader might disagree, and I mean that. I am willing to be instructed if anyone can show me photographic evidence of two identical dials. It is impossible for our memories to store visual details of dials and do it correctly. Therefore while we might *think* we have seen the twin of a certain dial before, the comparison of photographs of each would probably show differences. In this connection I have collected many photographs of white dial clocks, and I have no two which are exactly alike. Naturally no man's experience can ever be comprehensive: I am quite ready to be proved wrong, but it must be with photographic evidence.

The following list of Birmingham dialmakers is, I believe, fairly complete. I compiled this list from directories for the following years: 1777; 1791–8; 1800; 1803; 1815; 1816–17; 1819–20; 1822; 1828; 1830; 1835–6; 1839; 1841–2; 1849–50; 1858; 1860; 1862; 1865; and 1880. (A less extensive list with some variations was published by Mr Clive Osborne of A. A. Osborne & Sons Ltd, clockmakers of London, in *Antiquarian Horology* in 1966.)

BIRMINGHAM DIALMAKERS

E. Alldridge	1860
Samuel Baker	1839–50
Samuel Baker & Son	1858

T. Baker	1839
Wm Baker	1822
Wm Batkin & Son	1803 (? printed dials)
J. Beach	1850–80
Chas Bingham	1816
T. Bingham	1860
F. Byrne	c 1780–90–c 1800–10
Jas Geo Cooke	1858–80
Ed Crow	1839
T. Crow	1849
Dugmore & Foster	1816
T. A. Eaves	1865
T. Eaves senior	1865
Thos & Jas Egginton	1841–2
Eli Felton	1815
Wm Finnemore	1815–22
Wm Finnemore	1850
Wm Finnemore & Son	1828–36
Wm & Geo Finnemore	1839–42
John Wright Fletcher	1841–50
S. Francis	1839
Wm Francis	1815–30
W. & R. Griffiths (successors to Finnemores)	1858–65
John Harden	1816
Josiah Hayes	1800–30
Josiah Hayes & Son	1835–6
Wm Higgins	1839
Hipkiss & Harold	c 1810–20
Thos Hitchins	1858–80
L. Hobson	c 1830
Howell & Cooke	1849–65
J(ohn) Jukes	1850–65
John Jukes	1839
Josiah Jukes	1839–42
Thos Keeling	1800–15
Thos Leighton & Sons	1880
Mansell & Allen	1865
Minshull	1816–30
Hen Moon	1880
Thos Neal	1791–1803 (? mostly watches)

Wm Nicholas	1793–1822 (also clockmaker)
A. & J. Osborne	1800–3
Thos/Ann Osborne	1778–98
Thos H. Osborne & Jas Wilson	1772–7
Edward Owen	1803–20
Owen & Price	1800
John Potter	1835
Hen Prince	1880
John Riding	1816
T. Robinson	1865
Rich Smith	1777
Solomon & Co	1815
Francis Chas Swinden	1835–50
Swinden & Sons	1858
J. Vann	1865
Walker & Hughes	1815–35
Geo Waterhouse	1818–20
Sam Westwood	1880
John Wilkes	1803–15
John Wilkes & Son	1822–30
Sam Wilkes	1835–50
Wilkes & Baker	1815–20
James Wilson	1778—died 1809
Robert Winn	1815–39
B. Wright & Co	c 1820?
Christopher Wright	1835–42

The following concerns are known to have actually stamped their names on dials or falseplates as:

F. Byrne, Birmingham
Finnemore, Birmingham
L. Hobson, Birmingham
J. Jukes, Birmingham
Osborne, Birmingham
Osborne & Wilson, Birmingham
Osborne's Manufactory, Birmingham
Owen
Walker & Finnemore, Birmingham, an unusual example of c 1810, perhaps before these men set up independently

Walker & Hughes, Birmingham
B. Wilkes & Co, Birmingham
S. Wilkes, Birmingham
Wilkes & Son, Birmingham
Wilson, Birmingham
Wright, Birmingham
B. Wright & Co, Birmingham

The monogram 'W.H.' or 'W. & H.' stamped into the rear of some dials represents Walker & Hughes.

A punch mark , rather like a Maltese cross, appears on some dials by Wilson.

A number '16' is recorded punched into the rear of some dials, also by Wilson. On the rear of some Wilson dials is a black-painted initial R on a white background and this can be a useful identification feature.

Other symbols will no doubt come to light with further investigation. The men listed above supplied dials, or dials-with-movements; but not, so far as is known, movements alone.

While Birmingham was the main centre of dialmaking, there were others. At Halifax, Yorkshire was the dialmaking and dialpainting concern of Whittaker & Shreeve. They worked together from about 1810 to 1817. In 1817, however, William Shreeve, artist and dial-painter, met an untimely death at the age of forty-three when varnish he was making accidentally spilled over him and caught fire. After the gruesome death of Shreeve, William Whittaker appears to have carried on alone until about 1840. Their falseplates are stamped 'Whittaker & Shreeve, Halifax'. Some are known stamped 'Wm. Whittaker', and also 'W.W.–Halifax'; and I have seen movements on other makers' clocks, punched with Whittaker's name, so evidently he also supplied movements to the trade. Whether this is the same William Whittaker who was still working in Halifax as a clockmaker in 1866, I cannot say for certain. Strangely enough, a Sara Whittaker was in business in nearby Leeds from 1834 to 1837 as 'painter, clock dial enameller and japanner'. Japanning was another term for this form of imitation enamelling. She must have been connected in some way with the Halifax man, but I do not know the connection.

Another provincial dialmaking concern was the firm of Beilby & Hawthorn, Newcastle upon Tyne. They appear in directories from 1801 to 1811 as watch-glass manufacturers and Lancashire tool warehouse. Exactly during what period they produced painted dials is not known, but falseplates exist bearing their name.

This stamping of different names on the same falseplate is a strange business, which I cannot fully understand. Just recently I have seen one with 'Osborne & Wilson' stamped on one side and 'Wilson' only stamped on the other. Could it be that a dialmaker coming into possession of another firm's falseplate decided to stamp his own name on the other side?

In my opinion the iron falseplate was a device invented and sold *attached to the dial*, purely on white dial clocks. Opinions seem to differ in this respect, however. My view is supported by the simple fact that one finds many a white dial impressed with the same maker's name as on its falseplate, thereby indicating that the two were sold together. In fact some white dials could not have been sold without the appropriate falseplate, since the date discs are sometimes fixed between dial and falseplate and *attached* to each, eg the dial by Juler of North Walsham (see p 81).

One school of thought is that falseplates were sold as separate items for use on white or brass dials. It is possible that some may have been sold separately, though I cannot see the point of it. I do not incline to the view that they were ever intended for use on brass dials. There had never been a need in the past for such a device on a brass dial, where there was no problem in attaching dial feet between movement and dial. There seems nothing to be gained by using a falseplate to attach a brass dial to its movement. I have never seen one used in this way that did not look very suspicious to me. A falseplate of course can serve as a very useful device to assist someone in fitting a dial (white or brass) to a wrong movement.

The view that falseplates were supplied with the movement and not with the dial was expressed by E. L. Edwardes in *The Grandfather*

Clock. It is reiterated by Anthony Bird in *English House Clocks*, and by Eric Bruton in *The Longcase Clock*. This view seems not to take account of these facts: many dials bear the same names as their falseplates; these names are of *dial* not *movement* makers (though I would not quibble against the possibility that an occasional dialmaker might also have made movements, especially after 1840); not all movements on falseplate dials are of the 'factory-made' type, but considerable numbers of earlier white dial clocks have movements bearing features indicative of the clockmaker's own handiwork, such as scoring and marking out wheel positions on the plates; some dials have the datewheel pivoted between dial and falseplate; movements appear to have been made or at the very least assembled by the clockmaker himself to fit on to ready-purchased dials even as late as the 1830s (see end of Chapter 7), while it is usually suggested that the dials were bought to fit on to pre-made movements; the meaning in the 1772 advertisement by Osborne & Wilson, when they say 'such Methods used as will enable the Clock Maker to fix them to the Movements', can hardly apply to anything but falseplates, especially as falseplate dials bearing their names are well known; and finally, no British brass dial clock of unquestionable originality seems to have been shown to have a falseplate fitting: the Birmingham dialmakers are not known to have made brass dials and therefore a Birmingham falseplate would immediately seem to be 'wrong' on a brass dial.

Another important point on which I feel these three writers may be mistaken is their implication that all movements on falseplate clocks must be 'factory-made' ones, since this carries with it an unfair suggestion of inferiority to those made (or assembled) by the clockmaker himself. I know of north-country clockmakers whose movements on their early white dial clocks are indistinguishable from those on their (usually earlier) brass dial ones and which also have distinctive features enabling one to recognise them as the handiwork of the makers concerned. Such movements appear on falseplate dials and there is no question of their having been made in Birmingham. Three-train musical clocks, scarce as they are, perhaps more frequently appear behind white dials than behind brass ones.

Surely no one would suggest that such musical movements were Birmingham-made, despite their being attached to Birmingham falseplate dials.

I would not go so far as to say that a falseplate dial would never have a factory-made movement attached to it. Nevertheless it would seem more likely that in very late clocks of the late 1840s and 1850s the *absence* of a falseplate is much more likely to be an indication that movement and dial were supplied together. After all, if one workshop were supplying dial and movement what need would there be for a falseplate, when standardised dial feet could be made to fit into standardised movement plates?

As with all hints on identification of objects, one must use the available facts with care. Information on dialmakers can be an extremely valuable reference guide to establishing the age of a clock. But one should take it along with other guiding features, not on its own as firm evidence. For instance, a falseplate on a clock could conceivably be a replacement, in which case it would have no bearing at all on the age of the clock. Similarly, a stamped date disc could be a replacement standing in for a damaged original one. It is very unlikely that a falseplate *will* be a replacement, as it is a sturdy object hardly likely to suffer damage; just the same, this possibility must be borne in mind. A date disc is more prone to damage and it is not unusual for a replacement one to be fitted. In the case of a replacement date disc (of the aperture date type), the colour of the white background will often vary in shade from the rest of the dial, and the style of numbering can vary too.

The dialmakers have left us a whole series of clues and guiding features which give us a completely independent field for identification. If you simply take the trouble to look, the information is usually there, and you can observe it without even getting your fingers dirty. You don't need to dismantle the clock: just look closely with the aid of a torch or a good light. The evidence, however, should be used alongside the other dating evidence you have available to you, ie the dial style; the facts on the maker's life; and the case.

CHAPTER 4

DIAL PROGRESSION: PERIOD ONE, 1770-1800

Among the earliest types of white dial clock must surely be the kind with an all-white dial and gold-painted corner decorations, which resemble the rococo spandrels of a brass dial clock. Two such clocks are illustrated (pp 17, 18), one by William Terry of Bedale in Yorkshire, the other by Caleb Nicholas of Birmingham. I cannot prove that these were of the very earliest type, but it would seem reasonable to assume so, since later developments indicate that, as a general rule, the more colour the later the date of the clock. The dialmakers themselves called them *white* dials, not coloured dials. On these two clocks the gold-painted corner decorations are obvious attempts to copy the traditional spandrel corners of a brass dial clock. This is not, however, to contradict my earlier remarks about the one-piece brass dial, where *engraved* corner decorations appear to copy white dial corner patterns.

The full potential of what an artist might do to a clock dial is not yet realised. Here he is working in one colour only. He cannot achieve very much when colour, which is the essential factor of his medium, is denied him. To express the situation that way, however, is to put the cart before the horse. In considering dials of this type, we must imagine that we have never yet seen a coloured dial. The white dial used by Terry and Nicholas has certain advantages over the brass dial, the principal one being its superb legibility and

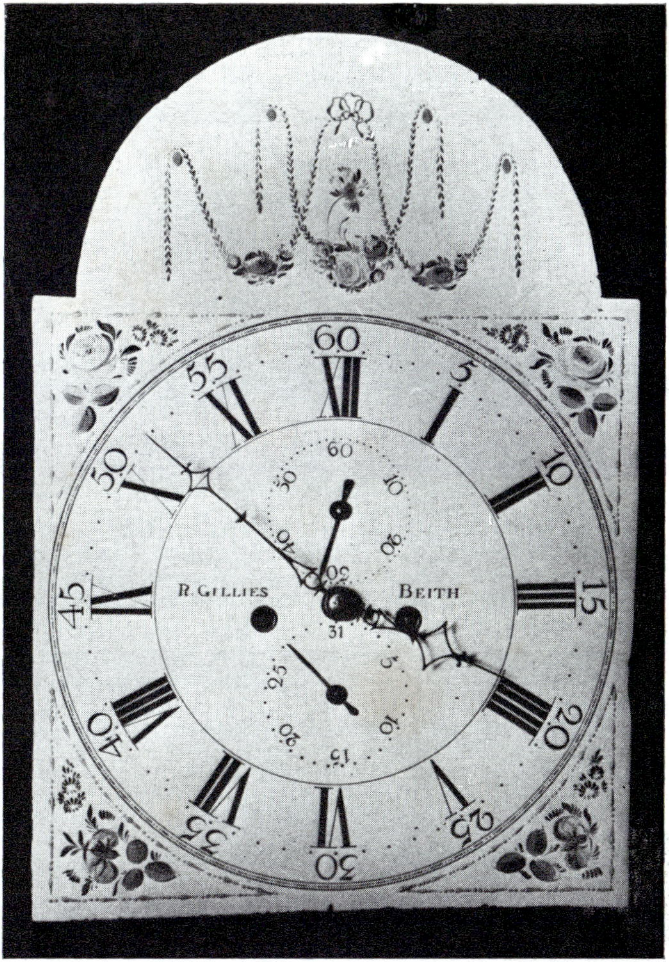

Page 53 (*top*) Robert Gillies, Beith, 1802; (*bottom*) Gillies signed and dated seatboard

Page 54 (*left*) John Fothergill, Knaresborough, c 1805; (*above*) Fothergill, c 1825

the second, its lower production cost. The prime function of a clock, of course, to which any other consideration is secondary, is its ability to tell one the time at a glance. This kind of dial is clearly legible, legible in a way which will never again be seen to be matched in later longcase clocks. A disadvantage of course is that the white dial is more prone to damage such as chipping.

The two examples illustrated are interesting in several ways. The Terry one is a typical country clock; the Nicholas one a town clock. I use these terms in their original meaning, not in the way that they are sometimes used, where 'town' means 'London'. By a country clock I mean one made for a country cottage or house (generally by a country maker). By a town clock I mean one made for a town house (generally by a town or city clockmaker). The former is a thirty-hour clock in a traditionally simple case. The latter is an eight-day clock in a more sophisticated case with a large amount of brasswork, all entirely original. The minute hand on the Terry clock is a replacement. The Terry clock is actually signed and the date inside is 1787, engraved by Terry into the brass frontplate. There is no name anywhere on it which might identify the dialmaker—not a falseplate dial. Terry is known to have worked at Bedale from at least 1770 to 1820 when he died.

The Nicholas clock was made about the same time, say 1785. It has a falseplate, but unfortunately one which is not stamped. The dial shows no impressed maker's mark. The date disc on this clock cannot help us with the dialmaker's name, as it is a replacement: I am certain about this, as I had it fitted, the original one being missing when I bought the clock. (A date disc on this type of clock is one which fits behind the pointer and by which the pointer is driven.) Known dates for Nicholas's working in Birmingham run from at least 1787 to 1808, but of course these are not terminal dates.

Apart from their all-white dials with gold-painted corners, these Terry and Nicholas clocks both have certain features in common with other clocks of the earliest white-dial period. The hands are of steel and non-matching. The original Terry minute hand would have been like the Nicholas one, or perhaps like the one by John

Day of Wakefield, Yorkshire, which I illustrate next (see p 18). Minute markings are indicated by *dots*, not by a continuous minute band. The minutes are numbered at intervals of 5, 10, 15, 20, etc. The hours are marked by *Roman* numbers. All these features are pointers which, when taken together, indicate the age of the clock. In the case of the arched dial a further indication is the largely undecorated arched part, bearing just the name. In later clocks, as we shall see, the arch is filled completely by an oil-painted scene.

While this type of dial was among the earliest styles of white dial, I do not mean to indicate that it was the only early style. Contemporary with the above type was another, a little more colourful but with many similar features, which I shall examine next. There are maybe as many dial patterns as there are clocks. One often sees very similar dials but never in my experience any two identical ones, a possible exception being the Terry type where decoration is so sparse that there can be little variety. Also of course it was not a case of changing in 1790 to style B, then again in 1800 to style C, and so on. Obviously changes take place gradually and designs overlap from different periods. This is one reason why it is always difficult in any work on antiques to give actual dates for styles, for there are exceptions to every rule. My aim in this book is to outline the normal, the usual; and it must be assumed that the exceptions by their very nature are less likely to be encountered.

This is the first book, so far as I am aware, which attempts to give serious guidance on the progression of and dating of white dial clocks. The reader may take it that before writing it I looked at what existing horological works say on the subject. If dates I suggest differ from those which any other writer may indicate, this is absolutely intentional; and as far as possible I shall give evidence to support my dates.

Now look at the other early type of white dial clock, as an example of which I have chosen a dial by John Day of Wakefield (see p 18). The lettering says 'I. Day', but at this time 'I' was often used for 'J', especially in 'John'. This clock was made about 1780–5. It differs from the first two only in one essential: coloured painted decoration in the arch and corners replaces the gold 'spandrels' (and the name-plate on the Nicholas clock). It is of course a thirty-hour clock and therefore we expect no seconds dial. Thirty-hour clocks very rarely have seconds dials, though on later examples they sometimes have dummy ones. Eight-day clocks on the other hand usually do have seconds dials.

The minute hand of the Day clock is different in design, being of a slightly earlier style than that of the Nicholas clock. The Day dial has two concentric black circles, which the Terry clock also has, but which do not show on the Nicholas one. This last feature is not, however, an important one, and I shall explain why. I know all three clocks: the Terry and Nicholas ones especially, as I have owned them. Over the years the black-painted numbering tends to wear off with cleaning and today one often has to restore some of this lettering. All three of these dials have been retouched in varying degree. The Nicholas one has not had its circles repainted, though they *were* there when it was new. I explain these small differences to avoid confusion, but they are really unimportant.

The Day clock, like the Terry one, shows the date through an aperture, not with a pointer as the Nicholas clock does. These early clocks more frequently do have aperture dates rather than pointer date-hands, but on eight-day clocks a pointer is sometimes used to balance, ie visually, the seconds pointer. The Day clock is typical in style of very many clocks, but on each similar one the corner and arch designs will vary. Sometimes the corners depict flower sprays, or fruit such as strawberries. The arch may show a bird, or a flower, or a spray of flowers, or fruit. The position of the man's name may vary as will the style of script; but the dial pattern, its overall outline, is the same. The pattern of numbering in particular is an important factor.

John Day is known to have worked at Wakefield from at least 1770 to 1801, in which year he died there. This dial has a falseplate stamped 'Wilson-Birmingham', which dates it as post-1778. A reasonable assessment would be 1780–5, maybe 1790.

The next dial illustrated (see p 35) is of an eight-day clock by John

Bancroft of Scarborough which at one time belonged to me. It was made about the same time as the Day clock, perhaps just a little later, say 1780–90. Apart from the arch, the dial is very similar to the Day clock, though of course it has the normal (with eight-day clocks) seconds dial and matching date pointer. The hands are of non-matching pattern in blued steel, the serpentine minute hand dating from just a little later than the straight one by Day. The corner decorations are enclosed by double ruled lines, which again may suggest a slightly later date, but this detail is not really significant. The arch carries a moon dial showing the lunar date (here it is 21) and the shape of the moon in the sky. As the month progresses the moon wanes by disappearing behind the right-hand 'hump', and as it fades away completely a new moon face begins to appear from the left. This type of moon dial is used for about fifty years or so from the start of white dial clocks; in exceptional instances longer. Those clocks without the moon are more common than those with such a dial, but when they have a moon dial this is the type they usually have, right through the period without changing. John Bancroft worked at Scarborough prior to 1807, and probably died before 1820.

Another type of moon dial appears on some square-dial clocks where a semicircular aperture below the numeral XII shows the moon phases. This type was often made around the turn of the century, about 1790–1820.

The dial by James Lomax of Blackburn (see p 35) is almost identical in style to the Bancroft one, differing from it in one respect only: the shape of the dial. This may give it a superficially different appearance, but it is not really an important difference, at least as far as stylistic dial progression and dating are concerned. Presumably a dial of such unusual shape had to be ordered and made specially for the customer in question. It is an extremely rare *shape* of dial, but the features on it are almost exactly the same as on the Bancroft one. The dial has a Wilson falseplate, which we know limits it to the period 1778–1809. James Lomax was born in 1749 and died in 1814, which does not help in narrowing down the period. My own estimate would be 1785–95.

The clocks by Day and Bancroft represent the stock types of dial (with or without moon disc) used during this early period of white dial production, 1770–1800, during which time there is not a great deal of change in the appearance of dials. The Day and Bancroft types of clock appear to begin contemporary with the Terry and Nicholas ones, but to have become much more numerous and, presumably therefore, more popular. Many clocks one sees have this type of dial and most would fall into the period 1770–90.

Our next dial (see p 36) is from an eight-day clock by Rowlands of Berwick on Tweed, Northumberland. Several men of this name worked there in overlapping periods so in this instance the maker's name is not a great help in dating. It is very often the case that, when a clockmaker signs with his surname only, it may well be a family concern (perhaps just father and son); when he works alone he usually adds his first name or initials. He does not 'sign' it with his own hand of course, but has the name painted on the dial for him by the dialmaker.

This clock by Rowlands was made about 1780–90. It has a false-plate by Osborne of Birmingham, and must therefore be pre-1803. The hands are of a non-matching steel pattern: the minute and seconds hands are both later replacements. Even where only the hour hand remains of the original two hands, it is apparent from the style of that hour hand as to whether the clock had matching or non-matching type. As we shall see later, the hour hand from a matching pair is considerably different in style.

The dial is similar to the one by Day: notice the familiar themes of birds and flowers. These must obviously have been favourite themes with the public of the day, bringing a touch of the beauty and colour of nature into the home. This is something the brass dial clock could never do, nor did it attempt to. 'Not only can we make more legible clocks more cheaply,' the white-dial makers must have claimed, 'we can give you something altogether more colourful and picturesque, bringing the brightness and colour of a spring day into your home all the year round.' True, when one remembers that these clocks were seen in candle-lit homes, which could be rather gloomy and drab, especially in the months of winter, with dreary days of rain and mist.

Today some people have a superstition about its being unlucky to have pictures of birds in the house. This cannot be a very old superstition, as vast numbers of these white dial clocks depict birds as a favourite theme. Probably the majority of the earlier-period white dials have as their theme either flowers or birds or both, to the extent that this in itself is almost a feature by which one can date a clock.

The next dial, by Forster of Sheerness, Kent (see p 36), shows nothing new. It illustrates the type of dial we have already examined as used on a southern clock. The strike/silent indicator in the arch is an interesting variation, and seems to have been a feature used much more often in the south than in the north. It is not really indicative of a particular period, though it would probably be unusual after 1830 or so. By rotating the pointer, one can 'switch off' the strike at night time or permanently if one wishes. A strike/silent dial was an optional extra and must have put up the initial cost of the clock slightly, by virtue of there being additional work in the clock-making. This clock was made about 1790. G. H. Baillie in *Watch-makers and Clockmakers of the World* lists the maker as a member of the London Clockmakers' Company and also gives this same approximate date. It is interesting that a maker with London connections should have bought his dials from Birmingham: the falseplate on this dial is stamped 'Wilson' of Birmingham.

Before going further it might be as well for me to mention the position of the square dial in relation to the arched one. The arched dial was always the more numerous type and presumably the most popular one. Square dials were used as well as arched ones from the very beginning of the white dial period. Generally, the square dial is more of a country or village product while the arched dial could apply to either town or country. The square dial was cheaper to make, if only because of there being less decoration to paint on, and the case hood too for the square dial clock must have been easier and therefore cheaper to make. While the square dial begins alongside the arched one, it does not survive for as long. Presumably its popularity waned in favour of the more colourful arched type, and/or dialmakers found it economical to stick to the one type of dial. The square dial is seldom used in the north later than about 1820 to 1830; in the south it survived a little later. The square dial was dying out about fifty years after its introduction; the arched dial had almost a further fifty years to run.

During the first thirty years of these clocks, from about 1770 to 1800, the basic dial styles remained more or less unchanged in so far as there was little progression during that time. With the opening years of the new century, however, things began to happen. As with almost all such evolutionary changes, the main thing to bear in mind is that throughout the period one of the most important factors influencing such progression was the saving of time in order to keep costs down and compete more strongly against rivals: rivals both in the sense of other clockmaking concerns and also other types and nationalities of clock. All the time then, changes were occurring partly to follow the whims of fashion, but more particularly where those whims could be exploited to save work, time and cost.

An interesting type of clock is one which has moving figures in the arch. On brass dial clocks one occasionally finds a figure of Father Time with his scythe swaying to and fro in the arch. This is basically a simple device, the moving figure being attached to an extension wire from the escapement arbor, so that as the pendulum swings to the right, the figure swings to the left, swaying once a second in other words. The dials which have moving figures are a little more complicated than ordinary ones, because of the construction of the purpose-made dial from composite pieces. The backcloth of the arch scene has to be set back from the foreground to allow a space wide enough for the support-piece of the swaying figure. These dials were more costly to buy. (For prices of dials, see Chapter 7.)

I have seen white dials with swaying figures which were made as early as 1800, though the majority seem to be later, about 1820–30. This is consistent, as we shall see later, with the general dial style which by this time has a fully painted scene in the arch. There are all sorts of interesting versions, being generally far more varied than the swaying figures on a brass dial clock. A common theme is a ship rocking on a painted sea, but all kinds of themes are possible. I have seen a swan swimming on a lake; a blacksmith striking his anvil; Eve picking an apple from the tree and then swaying across to hand it to Adam. Sometimes one sees two moving figures, eg two men sawing wood with a two-handled saw. These moving figures, not particularly common on northern clocks, seem to have been most popular in the west country. Perhaps a rocking ship was more appropriate for a seafaring county such as Devon.

CHAPTER 5

DIAL PROGRESSION: PERIOD TWO, 1800-1830

The period beginning about 1800 and lasting at most thirty years is one partly of experiment with dial styles. It is a hard period to define clearly and precisely, but there are a few factors which crop up regularly, if not constantly, during this time.

The first dial we show (see p 53) is from a clock by Robert Gillies of Beith, near Glasgow. It was made in 1802. Under the seatboard is written, presumably by Gillies himself, 'Robt. Gillies, Clock Maker, July 1802'. This is not proof, of course, of his having written it. It *could* have been written by anyone at any time. However, it is in a script of this period, and as the dial features suggest the same time, I am inclined to accept it as accurate. Gillies is known to have been working there by 1780.

With this dial we are just entering the new century, and much of the style of it is reminiscent of the old one. Minutes are marked by dots and numbered fully; the arch is not yet filled completely with painting; the corners are not filled but have flowery tracery only. There is no falseplate on this dial, and no dialmaker's name; but painted large on the back of the dial is the number 44, the meaning of which is now lost to us. One quite often sees numbers of this sort painted on dial backs. Perhaps they meant something to the dialpainter: maybe the number of dials he had painted. It could even have been the number of the clock the maker sold, though this is unlikely. It may have had some temporary meaning only. I remember once pointing out the numbering 1140 on the back of a dial which produced the suggestion that perhaps this was the time the dialpainter went for lunch!

What is new about this dial then? Just the hands. These are now of matching pattern, though still in steel, of course. Hour and minute hands match each other; seconds and date hands also match each other, but not the two larger ones. This is an arrangement which is normal from now on. Naturally, datelines must not be taken too rigidly. It is not impossible for a clock to have matching steel hands by 1795, or even 1790. But one can be reasonably sure that they would not appear by 1780. A line has to be drawn somewhere, and 1800 is a good clear date after which one would normally expect a clock to have matching steel hands.

The next dial of this period (see p 54) is one which exhibits several interesting transitional features. Another reason why I have used it is that, in the arch, on the stone against which the gentleman in naval uniform is leaning, 'Horace Nelson' is written. This does not show in the photograph, as the dial is obviously in a very shabby state. However it is reasonable to assume, especially since the dial style is also consistent with that assumption, that it was made about the time of Nelson's death, 1805. Fothergills were established in Knaresborough, Yorkshire, prior to 1807.

Some of the changes taking place in dial styles as shown by this example are as follows. The minutes are no longer numbered 5, 10, 15, 20, etc, but merely 15, 30, 45, 60, the other numerals being replaced by little star motifs. Minute numbering will soon be dropped completely, and this is a step on the way. The minutes are still marked by dots, but this too is soon to change. One factor which may have a bearing on the dropping of the five-minute numerals is the general increase in education. More people now could read and write, though by no means all, and with the increase in literacy and 'scholarship' of the average person comes a decrease in the need to spell everything out for him. People now knew from infancy that two past eight meant ten minutes past eight. This cannot have been the case when a clock was the possession of the few. The white dial clock was rapidly becoming cheap enough to be bought by the masses. Children growing up in a home with a clock in it soon learned to tell the time, while it can hardly have been as important to them to learn how to do this in a house where there was no clock.

As the minute numbering gradually drops out, along with this comes a switch in the hour numerals. Roman hour numerals, used on all earlier clocks, now begin to be replaced by Arabic numbers, as on this Fothergill dial. There may have been a slight saving in time in that these may have been somewhat quicker to paint than Roman numerals.

The arch painting is spreading. No longer a simple spray of flowers but a whole vignette scene takes up much of the arch, with a spray at either side. The painting of the corner decorations is also spreading. Again, this becomes more than just a spray and begins to fill the whole of each corner. The corner designs now often cease to be floral and become much more geometrical, the quarter circle or fan pattern, as illustrated here, being much favoured. The hands, as with the last example, are in matching pairs, still made of blued steel.

The nineteenth century is getting under way: things are moving. The changes mentioned above mostly arise from a saving of work. An increase in the *painted* area perhaps arose from a desire for a more colourful, showy dial; but the extra painting time is offset by a saving in lettering time. The features of this Fothergill dial are typical of the time, though of course some dials still follow the traditional pattern more closely (as with the Gillies one) or depart from it in more gradual stages, with a mixture of old and new styles in this changeover period.

The next dial illustrated (see p 71) is by John Palliser of Thirsk, Yorkshire, and was made in 1817. There is no guesswork about this. We know exactly, as the movement is also engraved with his name and the date '8th March 1817'. Why it has the precise date in this way, I do not know. I am only thankful to have found that it has, since dated longcase clocks are very uncommon. They are especially valuable for noting changes such as those we are discussing. This dial shows a mixture of features. The arch is filling up, though is not quite full. The corner decorations are of a semigeometric nature. The hands are of matching steel pattern. Then—a rather unusual feature—the minutes are marked by dashes, not dots, almost joining the black circles into a minute band, something which will in fact take place before much longer. The minutes are not numbered at all. The hours have retained Roman numerals. This is an early clock of John Palliser's, just as the clock by Fothergill was an early one of his production. The Palliser dial has no falseplate, the movement being fixed direct to the dial.

I shall now show you a later dial by each of these men (see pp 71, 54). Both have matching steel hands still; and the arch part of the dial is now entirely filled with painting. Both have full minute bands with no minute numbering at all; and the corners are completely filled with decorative painting, the dial by Palliser being of a geometric nature still. The Palliser dial retains the Arabic hour numerals of the transitional period; the Fothergill one has gone a stage further in a *return* to Roman numerals for the hours. Incidentally, the Fothergill dial is a 'dished' one, ie the entire centre area bearing numbers is in the form of a raised convex bowl, which is not apparent from the photograph. 'Dished' dials of this type are not uncommon from this time on. The second Fothergill dial has no falseplate, but fits direct to the movement. It has no dialmaker's name, but behind the arch, impressed into the dial sheet, are the initials 'W. H.' This must surely represent the firm of Walker & Hughes which, as we have seen from the list of Birmingham dialmakers, was working from 1815 onwards. The second Palliser dial is about 1820–5; the second Fothergill one about 1825–30. In these four dials we see the departure from the early type; experimentation through the early years of the century; and the return with the later Fothergill dial to a more standardised, but new, type. These styles are in *northern* clocks: not quite the same thing applies in the south, as we shall see.

Next we examine the dial (and hood) of a clock by John Farrer of Pontefract, Yorkshire, included partly because of its style and partly because it is a rather attractive clock (see p 72). It has Arabic hour numerals, no minute numbers, geometric corner patterns, and arch painting enclosed in an unusual octagonal box instead of the more normal oval shape, and matching hands now in *brass*. It was made about 1815–20. Hands in brass are a new feature which are shortly to become standard in all later longcase clocks (of northern or midland areas). Notice that their shape is not unlike those of the first Fothergill clock, and quite different from the later matching ones in brass. The Farrer clock has brass hands but of the normal steel hands shape.

Notice that this dial *appears* not to have a full minute band, the minutes being marked by dashes. In fact, this is an omission in the relettering, and when one examines the dial closely, it is obvious that it did originally have a full minute band, just like the second Fothergill and second Palliser dials. No factual data can be assembled on this maker, since there was a whole family of this name making clocks in Pontefract and the data on them are too confused to be of assistance.

Now we examine a dial by Woolston Roberts of Derby (see p 72). It has the traditional and still typical moon dial. By this I mean that where white dial clocks have a moon dial, this is usually the kind they have. Some square dials have a semicircular moon dial set in below the numeral XII, but this type is less common and is largely northern. For dating purposes a moon dial is disappointing since the same type of moon dial is used virtually unchanged over a long period, and it takes up all the arched top area, which would otherwise provide very helpful information from its decorative style.

One gets used to seeing moon dials like this one, as it is the commonest type. But in fact probably no more than 10 per cent of white dial clocks have these moon dials, and since they involved more work and cost in manufacture, presumably we ought to value a clock with a moon dial a little more highly than one without. They appear on thirty-hour clocks almost as frequently as on eight-day ones.

The two hemispheres below the moon disc are painted here, as normally, to represent terrestrial globes. Of course, they serve no purpose other than that of decoration. Presumably they arose from some dialmaker's idea of how to use up two almost circular spare shapes. They are usually, but not always, mapped out as a globe.

This dial by Roberts has corner decorations which are unique in my experience. The little oval urn decorations are half-round (convex) beads actually stuck on, and the painting is applied on them and around them in the usual way. The owner of the clock told me that at one time one of the beads fell off and was lost, and he replaced it with an imitation one, repainting it to match the others.

The dial has suffered a little with time. In particular, the date disc which shows through the aperture above the maker's name, is an incorrect replacement. Originally it was made for a brass dial clock and is of course made of brass. The seconds hand is a modern replacement. The main hands are apparently original, being of matching brass pattern.

There is a similarity in numbering on this dial to that of the first Fothergill clock, ie the 5 and 10 minutes numbers are replaced by a decorative motif. It also has Arabic hour numbers. A difference, however, is the full minute band: the Fothergill dial had dots. There is no dialmaker's name. The reference books immortalise Roberts, like so many other makers of white dial clocks, by failing to mention him at all. So here you will have to rely purely on my dating—about 1820. Why not nearer 1800? Well, the use of the full minute band and the brass instead of steel hands would suggest to me no earlier than 1815.

The next dial illustrated (see p 81) is by Juler of North Walsham, Norfolk, and it is a little unusual. All the painting on the dial is in gold with black outlines—different and very effective.

There is no corner decoration; in fact there is no decoration at all, apart from the gold scrollwork around the name. The dial is 12in wide. The use of just the name in the arch is reminiscent of the Nicholas clock. The steel hour and minute hands are probably replacements and can be disregarded for dating purposes. The steel seconds and date hands are probably original.

On the reverse of the dial there are a few items scratched on by a clock cleaner or repairer who must have serviced the clock in the past. This is an interesting feature which many clocks have, and of course you will see this sort of thing only if you look specially for it. Such things add to the interest of a clock, but you have to know what to expect in order to be able to watch out for them. The repairer's name could not be deciphered. Following it are these dates scratched into the black backing paint: February 72; February 74; January 77; April 83.

If we look at G. H. Baillie's *Watchmakers and Clockmakers of the World* we find: 'Juler, John, N. Walsham 1753–73'. With this to guide us we might arrive at a date of pre-1772, say 1770. If so we would be quite wrong.

The original hands would almost certainly have been matching steel ones, though you have only my word for that. The largely white dial is misleading, looking superficially like the Nicholas type, but not when you examine it properly. Arabic hour numbers, full minute band without numbering of minutes—these together ought to date the clock about 1810 or 1820, no earlier. The date pointer instead of the aperture date is often a late sign. An important clue is that the dial has a falseplate stamped 'Finnemore—Birmingham', and Finnemores worked there after 1815. The cleaners' dates of '72', etc, must be 1872 not 1772. The clock was made about 1820.

But what about Baillie's dates? Well, I say again: judge for yourself on evidence, then turn to a reference book for confirmation if you wish. But don't change your ideas because of a date given in a book. I am not saying that Baillie is wrong about John Juler, who may well have been working at the dates he quotes. But if so, he did not make this clock. This must have been sold by a later man of the same name, perhaps a son, unless John was still selling clocks in his nineties. The fact that Baillie does not show a John Juler working about 1820 does not mean that there was no such person.

Dials of this almost-all-white style seem to have been popular in East Anglia rather than throughout the country as a whole.

Finally, in this same period still, I illustrate a southern clock (see p 82) by Joseph Keys of Exeter. This dial, as is obvious, is in rather poor condition, the white background being crazed. This is a fault which affects some clocks as they age, though fortunately not too many of them. The top right-hand corner has been cleaned and re-lettered to show clearly the original numbering, and also to illustrate how dirty and faded the rest of the dial is. It is 11in square and has an aperture date. At first sight this may seem to be an early clock. But southern dial styles tend to be conservative and lag behind northern ones in some respects. Therefore we get used to adding a few years on to the *painting* style of a similar northern clock, with which we might compare it.

The numbering style should give the main clue: full minute band, Arabic hour numbers, no minute numbering. These suggest a date of post-1815. The matching steel (original) hands suggest a date of post-1800. By 1830 we shall see that further changes have developed. We are reasonably certain by these features to be in the period 1815–30. There is no falseplate on this clock, but careful examination showed 'Walker & Hughes' stamped into the rear of the date wheel. On the reverse of the dial is an unclear impressed mark reading 'W. & ?', which must have been 'W. & H.' Walker and Hughes we already know were Birmingham dialmakers working from 1815 to 1835, being consistent with the period we have already estimated. We said 1815 as an 'at least' date. A fair assessment would be, say, 1820–30. The case, in solid mahogany, has *branded* into the back-board and into the back of the door, 'J. Keys'; and penned alongside the name is 'April 15 1825', unquestionably the year in which the clock was made. Since I wrote this, the clock has been cleaned, and the same date and name were found engraved *inside* the movement.

Incidentally, Joseph Keys appears in no reference book. Not because he was not a clockmaker, but just because reference books are not 100 per cent complete. I therefore tried to look him up through local directories. He was a watchmaker at West Street, Exeter, from 1830 to 1837. By 1852 he was at James Street, and was still listed there in 1857. If therefore he did appear in a reference book, the book could well give his dates as 1830–57: rather misleading. However, while I was unable to locate a pre-1830 directory, we know

Page 71 (*left*) John Palliser, Thirsk, 1817; (*above*) Palliser, c 1820

Page 72 (*left*) John Farrer, Pontefract, c 1820; (*above*) Woolston Roberts, Derby, c 1820

from his thoughtfulness in dating this clock, that he was working prior to 1825. And, most important, we know from the dial style that the clock was made before 1830 and could not possibly have been made anywhere near as late as 1857.

CHAPTER 6

DIAL PROGRESSION: PERIOD THREE, 1830-1870

Having begun the previous chapter with a Scottish clock, it seems only right to begin this one with a Welsh clock, though perhaps it is slightly out of sequence here. I illustrate a clock (see p 81) by Stephens & Davies of Neath, Glamorgan. This has a falseplate dial, the falseplate being stamped 'Finnemore—Birmingham'. We know that this puts the date as post-1815, but as this concern worked through a long period, this is as far as the dialmaker's name helps in dating.

The dial is not unlike the one shown a little later (see p 83) by Richard Snow of Pateley Bridge, Yorkshire, though an eight-day one. When we compare the cases of each later, it is only fair to point out that Snow put considerably better cases on to his eight-day clocks than on to his thirty-hour ones. The arch area is entirely filled with painting, joining up to the corner paintings, which run into each other. Actually, this clock has five oil paintings which run into each other around the dial. A romanticised scene of a shepherd with his dog fills the arch. This liberal use of paint dates the dial as post-1830. So too do the brass hands, full minute band, Roman hour numerals, and the absence of minute numbering. I would put this clock at 1840, but 1830–40 would be an acceptable estimate.

The dial is fairly typical. One might normally expect a date pointer at this period; but, as here, an aperture date is not impossible.

The black numbering seems unduly heavy on this dial, but that is how it was originally. It has not been relettered.

The dial by William Hooker of Lewes, Sussex, is of course a southern clock. At first sight one might be deceived into thinking that the date of it is earlier than it really is. The dial is 11in square. Fortunately it bears an additional name, 'C. Waldrett', and the date '1833'. Dated clocks are very unusual, though of course I have tried to show a number of them in this book, one with a known date being by far the best for reference purposes. Don't expect to come across many dated clocks in the normal way of things, however, as I should think you would find a date on less than one in a hundred.

We can only assume that C. Waldrett was the clock owner who wished to have his name on the dial, and the date was perhaps to mark some special event. Clocks were quite often made as wedding presents, and of those that do bear the names of owners they often have the name of both the bride and bridegroom, as well as the wedding date. This clock was probably made in 1833 to mark a special occasion for the owner. An alternative explanation of the second name and date is given a little later on in this book. It is just possible that 'C. Waldrett 1833' could have been painted on the dial at a later date, but the dial-numbering style is consistent with this post-1830 period. I keep trying to anticipate possible objections on the grounds of something being a later modification or alteration, but in reality it is the exception which has been tampered with and one does not expect this to apply to the general run of white dial clocks one meets with. On the other hand it is best to approach a clock suspiciously and critically if you are thinking of buying it or even of simply studying it. If you are looking for faults you are perhaps more likely to spot them, but some people may look so searchingly as to find faults that are not really there. It is I think true that of all antique objects which may suffer from alteration, modification and sheer faking, white dial longcase clocks are among those *least* subject to such practices, probably on account of their value being until quite recently very low. A man with the talents of faking would not waste them on a thirty-hour clock by William Hooker, when he could be faking much more valuable stuff. The more modest the clock the more likely it is that no one has even thought of tampering with it.

In the north in 1833 they were generally using much bigger dials, and even a square one by then would measure 12in or 13in. The 'north' in this context and throughout most of this book means anywhere which is not the south: in other words, Wales, the Midlands and all central England would come under the heading 'the north'. East Anglia and any counties to the south of London, I would regard as 'the south'. Square dials seem to be less common in the north by this time, and an arched dial of 1833 in the north would usually be 13in or 14in wide, and even 15in is not unusual. White dial clocks were always more popular and therefore more numerous in the north than in the south, and southern clocks therefore are more likely to lag behind in fashion. There was not the same enthusiastic demand for them in the south, and I think it would be very uncommon to find southern clocks as large or as late as the northern ones. Clocks such as the J. C. Elliott one illustrated later (see p 93) would be virtually unknown in the south, which is why some of the old-fashioned clock books refer to any big clock as a 'Yorkshire clock', regardless of where it was made.

The Hooker clock is evidence of this fashion lag. The corner decorations of the fan type would be far more likely to occur on a northern or midlands clock about 1810; and by 1830 they are definitely out of favour in those areas, as will be seen in the examples I show. What reveals the age apart from the date on the dial, is the full minute band and the lack of minute numbering; also the return to *Roman* hour numerals which we saw in the Welsh clock. The case of this clock I cannot comment on, as I bought the clock without a case—and sold it the same way, I had better add. The hands too were lost. There was no falseplate on this clock; the long dial feet fitted direct to the movement. Nor was there any dialmaker's name.

I have learnt from a gentleman who owns an eight-day Hooker of Lewes clock, of the all-white-dial type, that William Hooker first appears in Lewes records in 1803 when his shop was at 12 High

Street. This eight-day clock is not unlike the Juler dial, but it has Roman hour numerals, a full minute band, and in the arch just a strike/silent lever—no coloured decoration at all. Hooker of Lewes is probably the same man listed by Baillie as being apprenticed in London in 1794. He would then be about fifteen years old, serving a seven-year apprenticeship, and so would be 'out of his time' about 1801. It would appear that he was making his presence felt at Lewes shortly afterwards. He appears in various directories up to 1832. In the next directory, of 1839, a Mrs Mary Hooker, clockmaker, is recorded at that address. Evidently William died before 1839 and not long after making the clock illustrated (see p 82). It is even possible that C. Waldrett was a journeyman clockmaker working for Mrs Hooker after her husband's death and that he was given 'second billing' on the dial.

The strike/silent lever I mention is a simple lever device which allows one to have the clock running without striking if one so wishes. It seems to be much more a southern feature than a northern one. I don't know why, but it is very uncommon on northern white dial clocks. A possible reason might be that northern clocks tend to have the arch fully used up by a painted scene, thereby leaving no space for such a lever which, when present, is almost always found in the arched area.

Another southern clock illustrated (see p 82) is by a Dawlish, Devonshire, maker, whose name unfortunately is so worn away as to be illegible. The 10½in square dial, aperture date, flower corner decorations—all these we would expect to be early features on a northern clock; but we know to expect them slightly later on southern clocks. The hands are of brass and matching pattern, as will apply to virtually all white dial clocks from now on. The hour hand of course has partly broken away. It has a full minute band without minute numbering: the minute band has been partly touched in to show the style. The hour numerals are Roman. All these later features give us a clue that the dial dates almost certainly as made after 1825, even later than 1830. If the maker's name had been legible we might have been able to look up some working dates to help. In this instance, however, we are lucky since the dial, though not a falseplate one, is stamped with the dialmaker's name. In the top left-hand corner (from behind) is stamped 'Wright—Birm.', and the same wording is also stamped into the reverse of the date wheel, ie the numbered date disc which shows through the aperture. We know that Christopher Wright of Birmingham does not appear in the records until the directory of 1835–6, and his name disappears after having been entered in the directory of 1842. We can therefore set the date of this dial as after 1830 to before 1849, and probably in the 1835–42 period. It is very satisfying to be able to pin it down to a short period in this way. There is no guesswork about it: we *know* when the clock was made. The case, we shall see, also *appears* at first sight to date considerably earlier than it is.

To those of us who are familiar with the established horological reference books, these last two clocks might come as a surprise. Almost without exception we are told that London led the field, and in dating a provincial clock we could expect to add twenty years or so to the date of that style in London. This may even be true in the pre-1750 period of longcase clockmaking. From the latter half of the eighteenth century onwards, however, the longcase clock was not really a popular form of clock in London, and it never became so firmly established there or elsewhere in the south as it did in the north. So far does this apply that, rather than being twenty years behind the southern fashions, the north was in fact twenty years ahead of them, as far as the development of the longcase white dial clock was concerned. William Hooker, being London-trained, presumably was as familiar with the fashions of the capital in clocks as anyone else in the southern provinces.

Clockmakers like the Snow family of Pateley Bridge, Yorkshire, were ahead of their southern contemporaries in catering for the demand for the newer styles of white dial clocks; and clocksellers in the midlands and north were prompt to anticipate the demand for clocks of ever increasing novelty, size, colourfulness, to the extent of their being leaders in the field.

The reason probably stems from the fact that, in the early nineteenth century, the Industrial Revolution centred around the coal-fields of the north and midlands, drawing other industries into those areas. Iron, steel, cotton, textiles, shipbuilding, engineering—these were the boom industries. The population migration was northwards, not to the south as it is today. The north was where the jobs, the prosperity, the clock market, and most other markets, were. This new opulence created a demand for larger, more extravagant clocks for the larger, more ostentatious homes of the newly rich of these industrial areas. But these clocks were not just for the rich. The newly not-so-poor-any-longer could also now afford longcase clocks as they too prospered with rising standards of living.

To say that the midlands and north were at the forefront of the new fashions in no way implies that these fashionable products were good. On the contrary, it could be argued that the more conservative styles of the south were an attempt at retaining pleasant and proven forms, while in pursuing novelty the midland and northern clock-makers sacrificed good taste and balanced design. I do not say that it is so, simply that it could be said to be so. I would be the last person to decry northern clocks, which developed into a style of their own. What view we take will depend on what type of clock we like personally, and this in turn will probably depend on what type we have grown up with and become used to.

What I am showing you is typical, not exceptional. These clocks represent what was then considered desirable and fashionable in the respective regions. Clockmakers, dialmakers and casemakers all made what was wanted, and they certainly knew the difference between northern and southern tastes. Part of our difficulty in forming an attitude to an antique clock today is that we judge it in the only way we can, that is by *our* standards of design and taste. We expect the clock not only to have survived intact and to function accurately, but also to conform to our tastes of a century or two later. This is really asking a lot! Yet we are unable to judge the clock in the same way as the contemporary customer did. How much of the furniture we make today will be considered tasteful a century or two from now, if it survives that long?

Whatever our own initial reaction to some of these clocks may be, one thing we can be sure of is that at the time these clocks were made they represented the height of fashion, each in its own way and in its own area. A family did not invest in a clock unless it was exactly the type they wanted. There were plenty of clockmakers, and customers had a wide choice. The typical clocks I illustrate represent those that the majority of people at that time considered not just acceptable but highly desirable.

The next dial (see p 83) is by Thomas Snow of Knaresborough, Yorkshire. It is considerably different from the Dawlish dial, yet I use it purposely since it is made by the same dialmakers, having a falseplate stamped with the name of Wrights of Birmingham, and also since it was made about the same time as the Dawlish clock. Snow's shop was in the High Street in Knaresborough between 1834 and 1844; Wrights appear in Birmingham from 1835 to 1842. This clock must have been made about 1835–40, and this should be apparent from the dial style alone. The full minute band is worn away, but it had one originally: I know this, as the clock belonged to me at one time. Roman numerals mark the hours; the minutes by this time are no longer numbered. The corners are filled with painting. The arch is filled with a complete scene. Hands are of matching brass pattern. What a totally different clock in regard to the painting, yet the basic features of numbering are similar to those of the Dawlish dial. This dial is much bigger, of course, being about 14in wide. The small pointer is a date hand, the date numbering being entirely worn off: if restored, the dial could look truly immaculate again.

There is an odd feature about this dial that may not be obvious at first sight. It looks like an eight-day clock with winding holes and winding squares protruding. It is in fact a thirty-hour clock with *dummy* winding holes, the squares blanked off behind the dial. This is a feature by no means rare on clocks. You bought your thirty-hour clock at the thirty-hour price, which was considerably less than the eight-day price. But your neighbours and visitors were deceived into thinking you had bought the more costly eight-day clock. Several such deceptions were practised. For instance, an eight-day

Page 81 (*left*) Juler, North Walsham, c 1820;
(*above*) Stephens & Davies, Neath, c 1835

Page 82 (*top*) William Hooker, Lewes, 1833; (*bottom left*) Joseph Keys, Exeter, 1825. The top right-hand corner has been cleaned; (*bottom right*) unknown Dawlish maker, c 1835

Page 83 (*left*) Thomas Snow, Knaresborough, c 1835; (*above*) Richard Snow, Pateley Bridge, c 1840

Page 84 Joseph Jefferies & Son, Leeds, c 1865

clock would generally have a seconds dial while a thirty-hour clock very rarely did. On some thirty-hour clocks, however, you do find a dummy seconds dial with a hand pinned to the dial. It never moved, it just completed the resemblance to an eight-day clock. Sometimes, too, one gets dummy date hands, which were never connected, but were there just for show. They are far less trouble mechanically, of course, than a real date hand, where if the mechanism is a bit worn or out of true it can often jam the clock. It seems to have been very common practice in the past to remove the operating pin and/or wheel connecting the date hand, and to leave the clock with the date feature not working, rather than run the risk of an incorrectly set date mechanism stopping the clock occasionally. On many clocks I buy this is an item which frequently needs replacing because of very shoddy repair in the past.

These dummy features are uncommon. If you do come across a clock with them, be rather careful. Sometimes one finds an eight-day dial which now has a thirty-hour movement attached to it; but it was never made that way. On such a converted clock the seconds and date pointers would probably not work, as there would be no linking wheelwork behind the dial. While a converted clock might still have winding holes, it probably would not have the dummy winding squares. The Snow clock is quite original, and so were many others with these dummy features.

I mentioned that thirty-hour clocks very rarely have true working seconds dials, and this is on account of the layout of the wheelwork (called the 'trains') of the clock. If a seconds hand were fitted on to a conventionally laid out thirty-hour clock, then the seconds hand would rotate in an anticlockwise direction, ie backwards. To reverse this, ie to make a seconds hand rotate clockwise on a thirty-hour clock, the maker would need to insert an extra wheel to change this direction of rotation. Some thirty-hour white dial clocks have this extra wheel and therefore a normally rotating seconds hand. Most of these thirty-hour clocks (white dial ones), with seconds dials, have dummy ones. I do know of one with a seconds hand which rotates anticlockwise. It was made by Joseph Belles of Fixby, near Huddersfield, Yorkshire. This clock, whilst a thirty-hour one, does have a falseplate dial, contrary to my earlier statements. This is because an eight-day dial has been used to give the superficial eight-day appearance.

The direction of rotation of a date hand need not be clockwise and commonly is not. The '*wrong*' direction of rotation is not so obvious with a date hand: one does not see it moving, because of the slowness of the date-change which takes place over a period of several hours usually.

After Thomas Snow we have a dial by Richard Snow of Pateley Bridge (see p 83). These men were brothers and second-generation clockmakers. Richard's clock is a little later than Thomas's, I would say. This is one of those clocks I mentioned as having a dummy seconds dial: the hand pinned flush to the dial and linked to nothing behind the dial. Richard worked in Pateley Bridge until at least 1837. A few differences indicate that this dial is a little later than the previous one illustrated. Typical corner scenes of churches, houses, etc, actually join up along the sides and run into each other. The arch painting runs down to meet the painting from the corners. Everything is colour and paint except for the white centre. The painting is crude but effective. The hands are of matching brass pattern. Notice the design of them: crowns on the main hands, a pattern normally found only on late clocks. It is not a common pattern but one sees it now and then, perhaps suggesting a date of 1837 or later, with the theme set from Queen Victoria's accession in that year? I would date this clock about 1840. It is a thirty-hour clock, very ordinary, and not a costly one—either when it was new or today. I have bought and sold dozens of similar ones, the dearest of which sold for far less than it would cost for a cabinetmaker today to make even the case.

There is something strange about this dial, something wrong, something that anyone who knows Yorkshire should spot instantly. The romanticised shepherdess is acceptable in this sheepfarming area. The bridges are quite pretty and appropriate to a place of that name, though, like the paintings on most dials, they are fictitious bridges and are not supposed to represent the bridge at Pateley. It is the cottages which are wrong. Presumably it is a Birmingham dial painted by a Birmingham man whose ideas of country cottages were based on the half-timbered, thatched-roof cottages of Warwickshire —quite out of place in Yorkshire. Cottages in Pateley Bridge bear no resemblance to these, which are not at all typical of Yorkshire buildings. There is no falseplate on this dial and no dialmaker's name to guide us; but it must be Birmingham made. This clock-maker sometimes used dials by Jukes of Birmingham (post-1839), whose name I have seen impressed into the reverse of such dials.

This paragraph explains why most dials have scenes which are *not* supposed to represent the places where the clocks were sold. An owner of a Leeds-made (or, more correctly, Leeds-sold) clock which might depict ruined abbeys in the dial corners, usually assumes that these are renderings of Kirkstall Abbey. They are not at all, of course, having been more than likely painted in Birmingham by a man who may not even have known that there was an abbey in Leeds. Ruined churches were typically favourite themes, and that is the reason for their being on clock dials, not because they represent the actual buildings of the locality. Very occasionally one may find a dial with a scene painted on and the place-name painted on below it as a title, but these are very exceptional.

Now we come to a late dial of a clock by John Catcheside Elliott of Leeds, made about 1855 (see p 93). Elliott came to Leeds in about 1850 and worked there for at least fifteen years. It is a superbly extravagant and exotic dial, beautifully painted. The paint completely fills the arch and the four corners run into each other. The corners depict Victorian ruins seen through luxuriant foliage. It is an unusual subject for an arch painting, and it may have been intended as some biblical scene. This dial and the next one (see p 84) are two of a whole series of dials of this type which suddenly flourish about this time. They show very bold, colourful and vigorous painting styles, a kind of wild last fling at decorative dials, often with much use of gold and silver paints and luminous paints. One sees them especially on Yorkshire-made clocks, but this may be partly because this was one of the few counties still producing them at this late date. This dial by Elliott has flowers in the corners, but they are totally different from the flower sprays of those more delicate early dials of the 1780s and 1790s.

Not a falseplate dial, this one: very few of them are at this time. It has the long dial feet fitting direct to the movement frontplate, probably indicating that Elliott bought in the whole clock ready-made, perhaps just assembling it himself. On the back of this dial is painted 'Pascoe No. 2'. Was Pascoe the dialpainter or dialmaker? I cannot tell. One often sees painted numbers or 'hieroglyphics' on the reverse of a dial, but the meaning of them is lost to us today. Quite possibly they may refer to the style of dial or the number done by that particular artist.

A point about this dial which may not be obvious in the photograph is that the background to the painted areas is coloured entirely in gold, like a gold backcloth to the scenes. This is a feature of many mid-nineteenth-century dials, with gold and silver and luminous paints highlighting certain details.

By this time the painting has progressed so far that the dial is almost one large, coloured panel of paint with a white circle within to show the numerals. The earliest dials were of course just the opposite, being almost all-white with just a little colour. Notice that we have no minute numbering at all; a full minute band; and large Roman numerals for the hours. Roman numerals never seem to have been used for seconds or date dials, perhaps because they are not so easily understood when they run higher than XX.

Matching brass hands are there, of course, at this period, with the seconds and date pointers also matching in brass. Once you have really looked at a dial of this sort, you cannot possibly confuse it with one of an earlier period, say of fifty years earlier. The two are totally different. It would be like confusing a car of 1920 design with one of 1970!

Another dial of a similar type to the Elliott one, is that by Joseph Jefferies & Son of Leeds (see p 84). In the arch is an attractively painted scene of the battle of Waterloo, the title 'Waterloo' being written just above the XII hour-number. Don't let the subject mislead you into thinking that the clock was made in 1815, the year of the battle. It was probably made in 1865, marking the fiftieth anniversary of that battle. Jefferies worked in Leeds from about 1850. The background to this painting, like the Elliott one, is all gold; and very effective use is made of luminous silver and gold paints for the uniforms and to highlight the church in the bottom right-hand corner. It has long dial feet fitting direct to the movement.

There is now no further progression beyond the Elliott clock. All the signs tell us that this clock is a late one, after 1840 certainly, maybe even after 1850, but there seems to be no further development later than that in the dial *style*. The Jefferies clock might have been made in the year of Wellington's death: 1852. I am more inclined to favour 1865, but then I saw the clockcase, which unfortunately has not been photographed. It was a large case, even larger than the Elliott one. The dial was most attractive, but the clock was hard to sell on account of its size. Like many others of its kind, it is now in the Netherlands.

CHAPTER 7

SUMMARY OF DIAL DEVELOPMENT

In this chapter I shall try to summarise briefly the main features in the progression of dial styles, which we have examined in the preceding chapters. On pp 96–7 these features are shown for quick reference in chart form.

Period One Dials begin basically with an all-white background. Numbering initially shows hours in Roman numerals and minutes in Arabic numerals. Each minute is marked by a dot. I describe the minutes as 'fully numbered', by which I mean that each five-minute period is marked in Arabic numbers—5, 10, 15, 20, etc. The painting at this time is usually sparingly applied. It can be simply in gold, showing gold corner 'spandrel' decorations, or it can be more colourful showing such typical themes in the arch and/or corners as a spray of flowers, a single flower (often a rose), a fruit (often a strawberry), a bird (usually one of the artist's invention rather than a portrayal of a real bird). Birds as a theme are usually colourful ones, and can sometimes be taken for eg a jay, or a bullfinch, or a bird of paradise, but they are usually not intended to represent an actual species. The painting is usually of objects rather than patterns. Moon dials are quite frequently found in the arch. Hands are of steel and do not match. If a calendar date is indicated on a dial it is usually by means of a disc showing from behind an aperture.

Period Two Experiment is the main feature of this period. Dial styles are gradually developing towards Period Three, but erratically. In the lettering the minute numbering is gradually reduced. The 5- and 10-minute markers begin to be replaced by a dash or symbol and minute numbering is often limited, still in Arabic numerals, to 15, 30, 45, 60. Hour numbers begin to change from Roman to Arabic. The marking of minutes by dots now gives way gradually to a full double band, each minute's portion being marked off along this band.

In painting the coloured area begins to take up more of the dial. The corners and arch begin to be more 'full', though perhaps not completely so. Fan patterns are popular as corner decorations, and so are various types of semigeometrical patterns such as sun-rays. Shells too have become a corner theme. The corner paintings are now more often patterns rather than representations of objects. The arch begins to fill often with an oval vignette scene, or the beginnings of a landscape or seascape. The arch begins to carry a picture rather than a pattern though this does not usually fill the arch panel completely. People appear both as individuals (eg Nelson) and as allegorical figures (eg Britannia; Justice; the Harvest, represented by a female figure holding a sheaf of corn; etc).

In the south and east of England dial styles lag behind the north and midlands in all except numbering. Dial sizes grow larger in the north, remain small in the south. Moon dials in the arch are not uncommon. Hands are still of steel but now match. Dates may be indicated by aperture dials or pointers; the former earlier, the latter later, as a general rule.

Period Three Numbering has now developed fully to a revival of Roman figures to mark the hours. Minutes, shown by a full minute band, cease to be numbered at all, as the 15-30-45-60 numbers of Period Two disappear completely. The Roman hour numerals now used are of course larger in size than in Period One, since they extend in length to cover the area previously taken up by minute numerals. This can mean, though not always, that Period Three hour numerals are heavy and clumsy.

The painted area continues to increase, particularly on northern-style dials. The arch is filled completely with coloured paint; so are the corners. Eventually, painting in the arch and corners joins up, forming a painted area within which a white circle remains. Religious scenes now often fill the arch, and landscapes and seascapes, rather than patterns. Hunting scenes are popular. Luminous colours are sometimes included. Favourite themes for the corners are ruins, churches, ivy-clad buildings, castles; also animals and hunting scenes. People appear both as specific heroes and as figures in a landscape. Ships are not uncommon. Southern dials lag way behind northern ones in style of painting, and do not seem to develop much beyond Period One and the beginnings of Period Two. Few clocks were made in the south at all in Period Three, relative to northern production. While southern clocks lag behind in *painting* styles, the numbering pattern on them is similar to that on northern dials. Moon dials in the arch are now less common. Hands match, but they are now normally made of brass and usually have much punched surface decoration to them, which the steel hands did not have. A date dial is often included, usually shown by a pointer, matching the seconds hand.

H. L. Honeyman began a most interesting examination of white dials as described in the *Proceedings of the Society of Antiquaries of Newcastle upon Tyne*, 28 November 1946. He classified the painting subjects into religious 10 per cent, patriotic 6½ per cent, romantic 16 per cent, sporting 10 per cent, landscapes 19½ per cent, flowers/birds 14 per cent, and so on. He is one of the very few people I know of who has even attempted to look analytically at white dials. However, by analysing them into subject matter, he was unable to arrive at any conclusions, but has presented an interesting list of 113 dial descriptions. What such an analysis did, without his realising it, was to begin to break down the clocks into periods. The religious subjects would mostly be painted after 1820, the birds/flowers mostly before 1800, and so on. What a pity that Honeyman did not research the dates of the makers whose clock-dial subjects he analysed; he would then have realised the importance of his recordings. The subject matter tends to be illustrative far more of the period of the clock than of the whims of the artist. It seems likely that clock dial-

Page 93 John Catcheside Elliott, Leeds, c 1855

Page 94 (*left*) James Lomax, Blackburn, c 1790. Note the architectural hood top; (*right*) John Bancroft, Scarborough, c 1785. Note arched top to door and square trunk corners

painters were painting the subjects which they were asked to produce by their employer, rather than being allowed to paint any subject they found appealing.

The notes given above may help to elucidate what is meant by the brief captions used in the chart on pp 96–7.

In concluding the section on dial development I would like to mention prices, as very little has ever been published concerning original prices, either of dials, movements, cases or complete clocks.

I have recently come across a fascinating document concerned with the transfer of stock in the year 1800 from a retiring northern clockmaker (Jonas Barber of Winster, Westmorland) to his successor. We can assume that the successor is taking over existing stock at cost (or maybe fractionally less) but the most interesting aspect is the price structure of white dials, which is worth setting out in full here. By this date, of course, brass dials were very much a thing of the past, and while some were still being made in the south, they were regarded as old-fashioned and out of favour throughout the midlands and north. The white dial prices were, in 1800 (words in brackets are mine):

Double Moon Clock Faces, 13in [arched moon dials]	£1 2s od
Land Skip 13in [landscape painted in arch]	16s od
30 hours, 13in with arches [arched dial]	14s od
13in square	12s od
12in with arches	14s od
12in square	10s od
11in square	8s od

It is interesting to note that the 'Double Moon' type (Bancroft type) cost nearly three times as much as the 11in square (Terry) type. Also of interest in this same document is the price of hands per pair (steel at this date) as too is the terminology used: 'Seven pair of Pointers at 7d per pair'.

There were sixteen dials altogether. Also in the transfer were four eight-day and seven thirty-hour movements, an interesting indication of the kind of stock carried by a full-time clockmaker, though one at the age of retirement. (This particular clockmaker was close

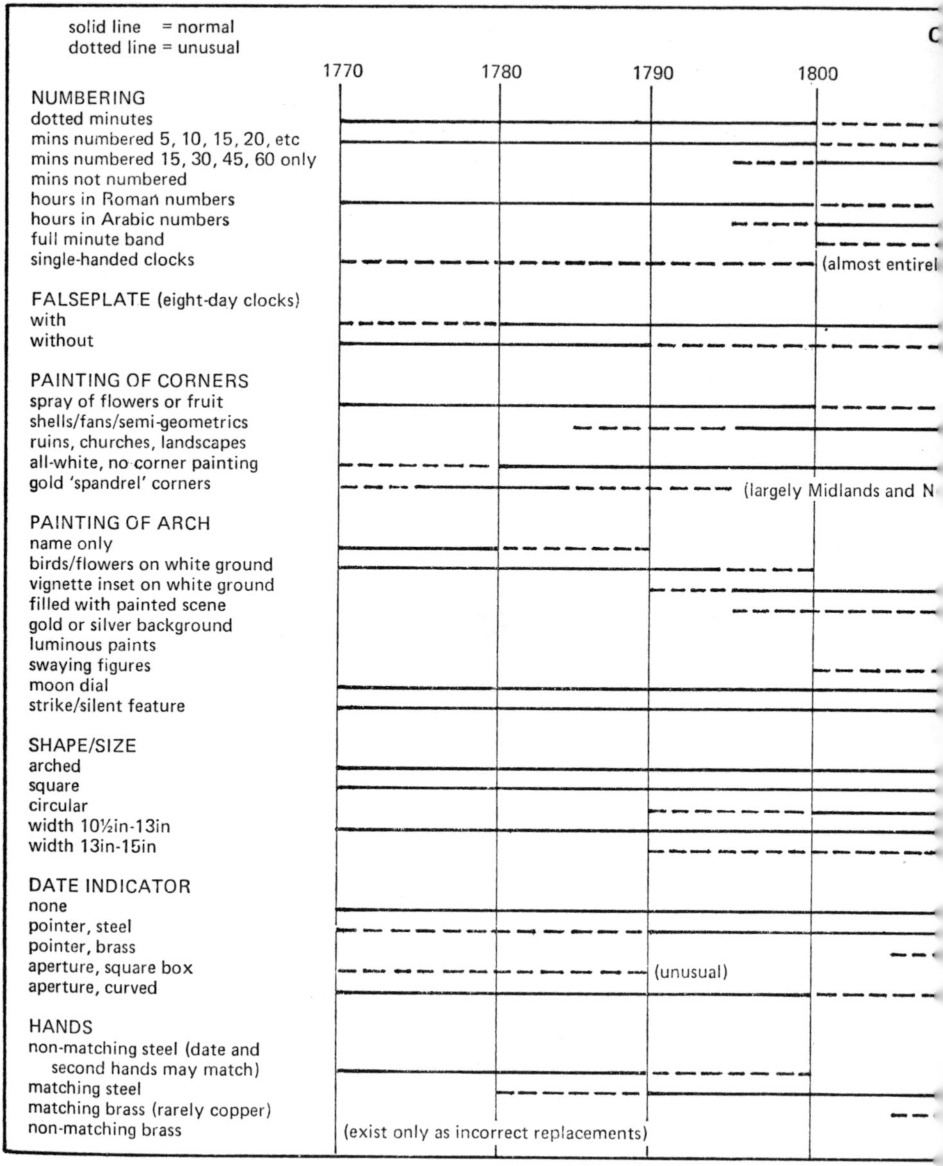
solid line = normal
dotted line = unusual
C
1770
1780
1790
1800
NUMBERING
dotted minutes
mins numbered 5, 10, 15, 20, etc
mins numbered 15, 30, 45, 60 only
mins not numbered
hours in Roman numbers
hours in Arabic numbers
full minute band
single-handed clocks
(almost entirel
FALSEPLATE (eight-day clocks)
with
without
PAINTING OF CORNERS
spray of flowers or fruit
shells/fans/semi-geometrics
ruins, churches, landscapes
all-white, no corner painting
gold 'spandrel' corners
(largely Midlands and N
PAINTING OF ARCH
name only
birds/flowers on white ground
vignette inset on white ground
filled with painted scene
gold or silver background
luminous paints
swaying figures
moon dial
strike/silent feature
SHAPE/SIZE
arched
square
circular
width 10½in-13in
width 13in-15in
DATE INDICATOR
none
pointer, steel
pointer, brass
aperture, square box
(unusual)
aperture, curved
HANDS
non-matching steel (date and second hands may match)
matching steel
matching brass (rarely copper)
non-matching brass
(exist only as incorrect replacements)

IAL FEATURES

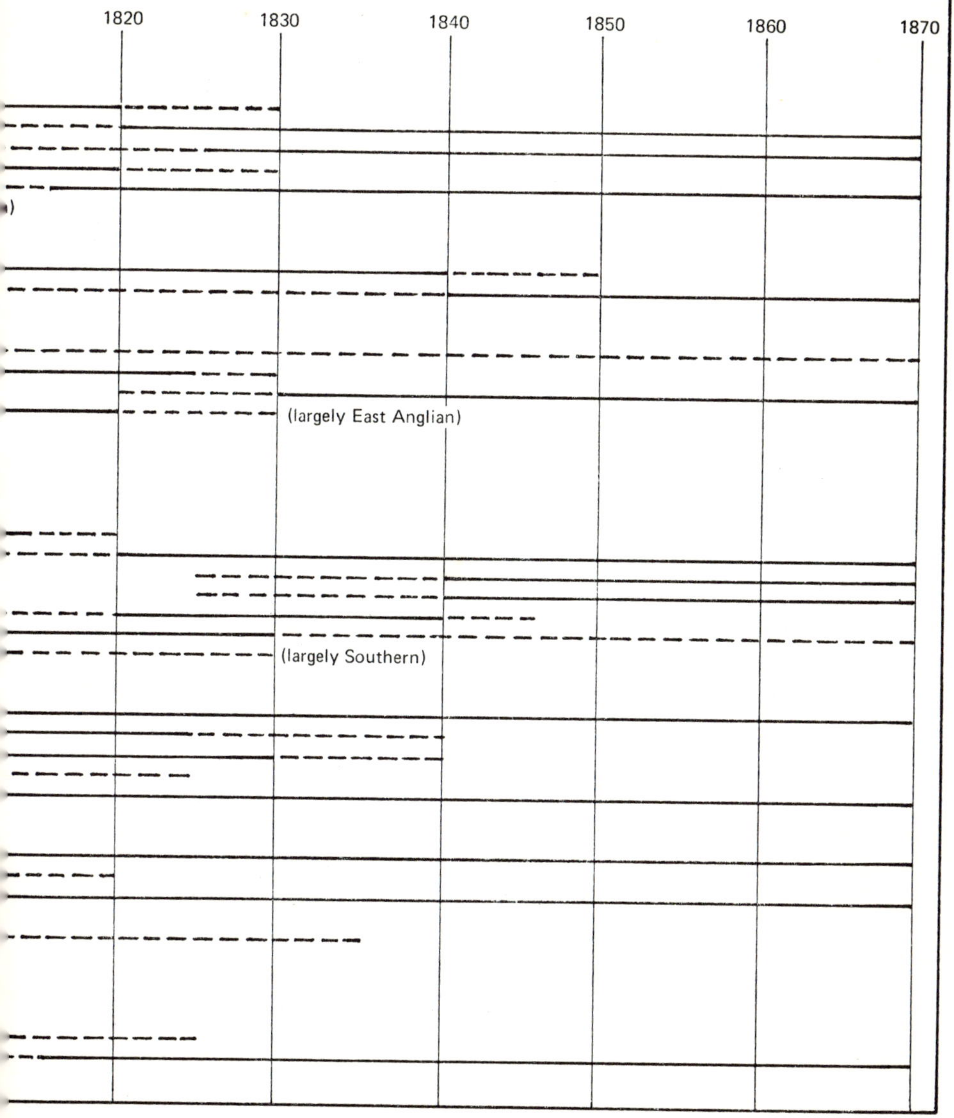

on eighty years old.) However, there is another most interesting point in this. He had sixteen dials but only eleven movements. This is supporting evidence for my view that the movements were made to fit on to dials, not the dials purchased to fit pre-made movements. It may also indicate that the maker could show his potential customers his 'range' of sixteen dials available, and make up the movement for whichever of them they selected, thereby eliminating the need to carry a large range of completed clocks.

Some very interesting records have been preserved concerning the clockmaker, John Manby, who set up shop in Skipton, Yorkshire, about 1815. Many invoices are preserved concerning the goods he bought, with sales ledgers of goods he sold, and these go back to about 1822. The family business still continues in Skipton under the same name, though nowadays as ironmongers. I am very much obliged to the Manby family for allowing me to examine some of these accounts, some extracts from which are published for the first time below.

John Manby, who would be typical of thousands of country clockmakers and hardwaremen of this period, bought in his dials from Birmingham, from Joseph May of St Paul's Square (from 1832 onwards), Messrs Mabson, Labron & Mabson, 'gunmakers' (from 1833 onwards), and also from John Balleny of 50 St Paul's Square (from 1837 onwards) . . . and no doubt from others too. These suppliers were mainly merchants supplying the hardware and clock trades with all kinds of wares. The dials from May's seem to have been almost half the usual price, but perhaps of poorer quality, as Manby bought from the other two suppliers at the same time despite having to pay them the full prices. It is interesting to compare these prices with those of Jonas Barber thirty years previously; Barber, incidentally, also bought at least some of his dials from Birmingham (from Osborne & Wilson).

Mabson's charged for '14 inch (ie wide) arch 30 hour dials, assorted patterns, no seconds dials, open month (ie aperture month dials) and named (ie with seller's name painted on)' . . . 15s 0d each. The explanations in brackets are mine. This was in 1833 and 1835.

Balleny charged in 1837 for:

14 inch arch 8 day dials, solid months, & seconds	16s od
14 inch ditto, 30 hour	15s od
14 inch ditto with convex* centres	16s od

However, May's much cheaper dials were in 1832 and 1835:

14 inch arch 30 hour	7s 6d	
14 inch 8 day	8s od	
14 inch 30 hr. single figures	7s od	[? hour numbering only]
14 inch 30 hr. ship to move	12s 6d	[rocking ship in arch]
14 inch 30 hr. Adam & Eve	14s 6d	[moving figures in arch]

Manby bought other clock fittings from these suppliers, some of the prices being (descriptions as used in accounts, my words in brackets):

spire clock balls [set of three]	1s 5d	[spire finials]
eagle clock balls [set of three]	1s 7d	[eagle finials]
clock hands [per pair]	8d	
seconds hands [each]	2½d	
clock case hinges [each]	4d	
2in clock roses [paterae] .... approx.	½d each	

Notice how much cheaper are these stamped brass items compared with the cost of cast ones as detailed on the Nicholas clock case of fifty years earlier. (The costs are given in Chapter 8.)

Manby bought his movements from Greaves & Newton, 'knife-makers', Portobello Works, Sheffield, not as complete ones but in ready-to-assemble form. This supports my view that falseplates were sold with the dials, and not with the movements. Falseplates often have 'Birmingham' stamped into them—none is recorded with 'Sheffield' stamped into it! Movements seem to have been bought by the weight of worked materials at so much per pound weight:

3 sets of 8 d(ay) clock, brass at 1s lb.
3 sets of forg'd work & pinions at 1s 2d

* See, for example, the dial by Fothergill illustrated on p 54.

There are a number of such entries, some more explicit, costing the brass parts and steel parts separately. Jonas Barber too had sold his (presumably already assembled) movements by the pound weight: 'Seven 30 hours movements 39½ pound @ 1s 4d per lb = £2 12s 8d'. This system of selling worked materials by weight may have been the standard method used in the trade, though I do not know of any reference book which mentions this practice.

Manby sold a 'new clock & balls' to a Mr Lister of Grassington, a joiner, for £4 2s od. We can assume that this clock was without a case, which Lister would no doubt make himself. Manby's charge for cleaning a clock, by the way, was a very reasonable 1s 6d! Gillow's, the Lancaster cabinetmakers, sometimes sold cases complete with clocks. One such clock they sold in 1771 for £4 4s od, plus a case for £3 13s 6d, presumably a brass dial eight-day clock. In 1799 they paid £2 11s od for a thirty-hour clock with 'china' dial from Newby's of Kendal. This would be a painted dial, of course, which Newby's probably got direct from Birmingham and fitted their own movement on to. This was to be sold by Gillow's in a case of their make. A Gillow case cost the customer from as little as £2 5s od in 1771 to perhaps as much as £10.

CHAPTER 8

CASES: PERIOD ONE, 1770-1800

One question which most people ask on first looking at a longcase clock is: 'Is it the original case?' This is a question to which one can very seldom give a definite 'Yes' in answer, as in most instances there is no certain way of knowing whether the clock has been re-cased at some time in the past, unless it has been put into a case of an obviously wrong period.

Very rarely one may find a casemaker's label somewhere on the case, perhaps at the back of it, or inside the door. I know a clock by Agar of York (a brass dial one in fact) with a paper pasted on to the back which reads: 'To Mr. Agar, York (? with) 2 clock cases'—the word I have put in brackets would appear to be 'with'. Another case by Woolston Roberts of Derby also has a delivery label on the back, the wording of which is given fully under the detailed description of the case. Neither of these labels, however, gives the name of the casemaker, being apparently just delivery instructions. These two instances I mention are very exceptional, and for the most part all one can say is that the case is *apparently* or *probably* original.

It is a fact that, in the past, clocks were often recased. If the original case became badly infected with woodworm or rot or suffered any one of the hundreds of mishaps which could happen to it, or even if the owner simply decided on a case of newer fashion, then in any of these instances a new case might have been acquired for the clock, either as a brand new one or as a second-hand one. This replacement case could be of the same age and style as the previous one, older, or newer, but usually it would be newer. A clock of 1770 recased in 1810 would most likely have been put into a new (1810) case. Cases were far more subject to deterioration than were clocks. Any dealer knows that it is fairly commonplace to come across a clock without a case, but the reverse (a case without a clock) is hard to find.

In this situation it is instantly obvious what is bound to happen. Suppose you are an antique dealer and come across a fine clock, say an early eighteenth-century one, with a case beyond repair. You then see an insignificant country clock with a 'suitable' case. You buy the country clock and rehouse your early clock in that case. The merits of such an action are debatable but, all the same, this was done in the past and is still done today. The biggest query, of course, is: 'What is a *suitable* case?' Ideally, a suitable case would be one of the same age, dial size, and type as the clock you are rehousing, but some dealers may have been prepared to accept a compromise or may not have had sufficient knowledge to determine what sort of case would be 'correct'.

Potential clockbuyers may now be wondering how, in the final analysis, they can know what they are buying. There is only one answer to that, and that is to know for yourself the various case styles and types, so that your own knowledge of the subject equals or surpasses that of the dealer you are buying from. I will try to give some hints to help with this, but first I must relate what happened only recently concerning the clock by Caleb Nicholas of Birmingham.

I was showing the clock to a man who was a clock enthusiast and a collector with considerable experience, but who said quite openly that he had never looked seriously at white dial clocks. He had been influenced by certain writers into thinking that white dial clocks were somehow not clocks at all. He did not look much at the dial, since with his preconceived idea, there was no way in which he could begin to actually like it. But he looked at the case, and expressed a measure of satisfaction with that. He then found that the attractiveness of the case conflicted with his ideas about white dial clocks. How could he actually find himself beginning to like a white dial clock, when all his reading had taught him that they were merely tasteless degenerations of the brass dial ones? The answer was not long in coming. 'Of course,' he said, 'a case as attractive as this one must have been made for a brass dial clock, and someone has later put this white dial clock into it.'

(Before going further, I may say that there is nothing magnificent about this case. It is mainly oak, with mahogany bandings and trimmings here and there. The brass fittings are of exceptional quality, and the proportions are good. It is a well-balanced and very nice case of its kind, but there is nothing really outstanding about it.)

I disagreed with the man politely and took off the hood to let him see that there were no obvious alterations to the seatboard and fixing of the clock.

He next looked at the movement. 'What sort of date would you put on that?' he asked. 'About 1820?' I said that I would date it about forty years earlier than 1820, and this quite took his breath away. This was probably because most antique objects tend to arouse in us a greater reverence the older they are. Now he was quite convinced that he was right and I was wrong. Then I mentioned that Baillie (G. H. Baillie, *Watchmakers and Clockmakers of the World*, known for short simply as 'Baillie') gave details for this maker as 1787 to 1808, not so far away from my own estimate of the clock's age. (I shall explain later why one does not take dates such as Baillie's as being precise limits.) The fact that Baillie gave a date similar to mine must have impressed this man, though one could not hope to convert him in the course of a few minutes' conversation.

This story is important in illustrating several points, mainly that the man was quite conditioned to the view that the white dial clock was the poor relation to the brass dial one and therefore, by his reasoning, would have had only a poor case. The fact alone that some white dial clocks are in splendid cases, despite many of the best ones having been taken to be put around brass dial movements over the years, tends to disprove his view. As to the dating of the clock, it was obvious that having never troubled to look at them, he could hardly expect to be able to date one.

The suggestion he made that someone had taken this case from a brass dial clock to put around the white dial one was quite ludicrous. In 99 per cent of instances of case-swapping precisely the *reverse* is done, ie a good case is stolen from a white dial clock to be put around any old brass one. That is where the financial gain lies and any dabbler in clocks who started to swap cases in the way this man suggested, would soon find he was on the shortest road to the bankruptcy court.

In this instance under discussion I was quite convinced that the case was the correct one. I bought the clock in a very dirty and shabby condition from having lain in a cellar, where it had not been touched or tampered with in very many years: one had only to see the state it was in to appreciate this. Forty or fifty years ago there was not the present heavy demand for clocks, and it could hardly have benefited anyone to have swapped this case. Most important of all was that the style of the case was right for the period of the clock.

A couple of other little hints might be worth mentioning to help you spot a swapped case. Firstly, the seatboard arrangement often has been altered and will betray more modern signs of sawing and chiselling than other parts of the case (though with little trouble new cuts can be made to look old with a touch of stain). Secondly, in most cases on the inside of the backboard there will be some sort of scratch-line caused by the pendulum-adjusting nut, below the bob, having rubbed against the back of the case when leaning slightly back from the vertical. If one puts a different clock into that case, its pendulum will not usually swing on quite the same level, and a second and newer scratch-mark might begin to appear. For more complicated reasons which I will not go into here, neither of these pointers is *proof* of a case change, but they are helpful indications if taken in conjunction with other factors.

Another point is worth mentioning in this connection. On many cases there is some sort of sign of wear caused by the descending weight(s) rubbing against the interior framing below the case door, sometimes to the extent that a couple of hollows are worn into the wood. If the descending weight(s) touch the interior framing (or fail to touch it) in different positions from these worn spots, this can be another indication of the movement's having initially belonged to a different case. On the other hand, repositioning of the attachment of the gutlines to the seatboard, as when new guts are fitted, can also affect the positioning of the falling weight(s). Similarly, if replacement weights are fitted, they could be of a different shape from the former ones and thereby hang differently as they come down. This pointer, then, is by no means an infallible guide, but it helps. When one sees a seatboard that has been tampered with, a pendulum that rubs in the wrong place, weights that descend in the wrong positions, all on the same clock . . . one ought to begin to feel suspicious.

One thing more I ought to mention, obvious though it may seem. A check should be made that the dial fits the case. Not infrequently I have seen the larger white dial clocks, say with 14in- or 15in-wide dials, in cases that were obviously designed to take 12in or 13in dials. A correct dial usually fits in such a way that the numbering, corner and arch decorations all show fully, and there are no gaps around the edge. When too large a dial is put into a case, the corner decorations are cut off by the door and part of the arch painting is obscured too. Dials were made larger as time went on, and what such a switch usually means is that a dealer has come by a good 12in to 13in case (with or without a poor clock in it) and come by a good *later* clock of the 14in or 15in size (with or without a poor case). When this happens not only is it apparent that the dial does not fit, but also that dial and case are of different periods. And so we come back to the same advice I offered you initially—get to know for yourself the styles of different periods. Then you are in a stronger position to make your own decision.

Let us now look closely at the first two cases, those of the William Terry and Caleb Nicholas clocks. Both of these I believe to be original. The Terry case is made of oak: plain straight-grain oak with a pine backboard. This was a farmer's clock, made in 1787 for a farmhouse or cottage, not for the lord of the manor. Made by a local joiner, it is simple but well balanced in outline. Neat and plain, it is just over 7ft in height. It has the original bracket-type feet (no plinths on north-country clocks as a rule) with two front feet only, no back feet. This is intentional, since it saves carpentry and the clock is far easier to stabilise on an uneven flagged floor when it has just the two feet, and not four. The back simply leaned against the wall; frequently the clocks were screwed or nailed to the wall. The hood pillars are simple wooden rods with turned capitals and the bases painted gilt to look like brass fittings. Turning was cheaper than brass and the joiner could no doubt do this himself and have the job completed before the fancy brass cappings could be ordered, let alone delivered, from the brassware merchants.

The case door is long and slender. Doors are made shorter as the years go by. Generally speaking, the longer the door (proportionate to the case), the earlier the clock. This case is strictly traditional in style, just the same as that made for a brass dial clock and typical of a simple country style made for twenty or thirty years either side of that date—1787 in fact. It is as small as the smallest London clocks (rare museum pieces excepted) and several feet smaller than many of them. This Terry case is typical of many thousands of provincial country cases.

'The Yorkshire clocks stand in a class by themselves. The cases, hardly without exception, are gigantic—almost elephantine . . .', so say H. Cescinsky and M. R. Webster in *English Domestic Clocks* in 1913. In making this totally incorrect statement, the writers go on to illustrate the point by showing photographs of three *Lancashire* clocks, which they misdate. We are forced to the conclusion that they had never seen a Yorkshire clock! Well, the Terry clock *is* a Yorkshire one, and as neat a little clock as you could wish for. It is absolutely typical of this style and period.

The Nicholas case is entirely different. It was made about the same time but for a quite different customer: a town customer with larger, taller rooms in his house. Such town clocks were made taller and grander for the larger houses—not mansions, not stately homes—but the ordinary town houses of more prosperous citizens.

This clock stands 7ft 8in to the top of the highest finial, or 7ft 2in without finials.

Look at the brassware on this case. All of it is original and in my experience quite exceptional. The brass caps and bases for the hood pillars at this period cost the casemakers between 5s od and 7s od a set; the quarter caps for tops and bottoms of the trunk quarter pillars cost perhaps 3s od to 4s od a set (as against 1s od or so for wooden ones). Add on about 7s od to 10s od for finials, 2s od or so for paterae (the brass discs on the faces of the horns), add about 8s 6d for gilding all these, and you arrive at a cost for brassware alone (without counting hinges, locks, fastenings, etc) of over £1 10s od. This was to add to a total cost for such a case of about £4 10s od to £5 10s od. In other words brass fittings such as on this clock would add about one-third to the cost of the case—the cost of *manufacture*, that is. Then the cabinetmaker adds his profit, and the clockseller too presumably adds his profit on the case as well as the clock. I base these figures on those shown in the business records of the cabinet-making firm of Gillow's of Lancaster, whose records include detailed costings of clock cases with sketches in some instances. (See *Gillow's Clock Cases* by Nicholas Goodison, published by the Antiquarian Horological Society, 1968.)

The brass fittings, then, add about a third to the original price of the Nicholas case. The price of this case would in any event be considerably higher than the Terry case as it has much more work in the shapings and bandings. The Nicholas case is by far the better in the sense of being a better-quality piece: better woods, better design, better workmanship, nicely proportioned. Today, however, with somewhat smaller rooms, the Terry clock would fit more comfortably into more homes and therefore might appeal to a wider market. The two, while contemporary, had a quite different appeal at the time they were made and still have today. You cannot compare one with the other except to illustrate their differences rather than their similarities.

The hood top of the Nicholas clock is finished with swan-necked pediments, sometimes referred to as *horns*. This method of finishing the hood top continues right through the white dial period and is perhaps the most popular style of all, especially on arched dials, though it is also used on square dials occasionally. It was especially popular in the north and midlands.

The door, like the Terry one, is as long as it could be, filling the entire length of the slim trunk. The figuring of the oak stands out well on this door. These figurings are called medullary rays and occur only in oak. I have read books which decry this medullary figuring as spoiling the look of the wood, but I personally quite like it. Whether people today like it or not, there can be little doubt that at the time this clock was made this sort of ray figuring was considered very desirable. You will see this figuring on the door, base panel and hood of the Nicholas clock. On the sides, however, there is not a single ray, the wood being absolutely plain and far less interesting. This happens on very many oak clock cases: the sides, which do not show up so importantly, are often quite plain; yet the front faces have rich ray lines. This is not by accident but by design. A similar thing happens with mahogany cases, the fine figurings being used for the faces and plain straight-grain for the sides. Sawing logs straight across produced straight-grained oak. One had to *quarter-cut* the wood to get the medullary rays to show, and this was a more wasteful, more time-consuming, and therefore most costly process, and also not quite as simple as it may sound. The Terry case is all straight sawn: easier, cheaper, less spectacular but functional.

A few other features of this case are worth mentioning since they are typical. Notice the mahogany banding around the edge of the trunk door, hood door and base—a feature often found in the 1770–1800 period. It begins before 1770 really, but in this book we are not concerned with pre-1770 clocks. Sometimes the banding is set in about 1in from the edge, which is perhaps a slightly later sign. This cross-banding is not done just for effect. Its real purpose is to present a cross-grain all round the outer edges for strength. Cross-grain wood is less likely to chip away or chafe in use than long-grain. Sometimes, but not often, long-grain banding is used. Of course long-banding is for nothing but show. It frequently chafes away at the exposed edge.

The shaped top to the door is another feature to notice. The flat-topped Terry door is the traditional shape, though sometimes an arched door top echoes the shape of an arched dial. By the 1770s door shapes start to develop. The Nicholas door shape is typical of this time; but watch for the exotic developments to come in the next century, when cabinetmakers begin to strain after novelty in shape.

The pillars of the hood are reeded, as are the quarter pillars of the trunk. Strictly speaking, these should be called pilasters. More work is involved in making these than with those of the Terry clock. Reeded pillars again are typical of the 1770–1800 period. Watch what happens to these slender reeded pillars in the next century, as casemakers continue with extravagant experimenting.

The bracket feet on this clock do not show up too well in the illustration, but notice that they *are* bracket feet, miniature versions of those on a chest of drawers of the same period. Some strange things happen to feet in the next century, that is when owners of clocks have refrained from sawing them off to get a tall clock into a low-ceilinged room.

These few pointers, I hope, will help you to fix this case in your mind. It is an excellent example of this type of clock because so many things about it are just right as typical features of the period. Fix those Corinthian brass capitals in your mind too, as you will seldom see their like again in the next century.

These first two cases exhibit the two quite distinct types of clock: country clocks and town clocks. As time goes by, this distinction becomes blurred and eventually even the village clocksellers were putting their clocks into more imposing *town* styles of case, as we shall see.

I shall enlarge on the point I touched on above when I mentioned sawing off the feet of a clock. On most clock cases the feet were the first thing to deteriorate. Many floors in the past were stoneflagged. People washed these by swilling water across them and brushing out the water and dirt together through the doorway. A few decades of that sort of treatment can soften up even tough English oak and make it an inviting home for woodworm, against which it would normally be almost immune. The damp wood starts to rot, the worm starts breeding, and before long the feet collapse under the strain; or they are sawn off and the process starts all over again, this time a little higher up the case.

If a clock has banding like that on the Nicholas one all around the hood and the door, but only on the top and sides of the base panel, then something is wrong. No designer in his right mind would do that intentionally. What has happened is that the case feet or plinth have been sawn off for some reason, and perhaps a new plinth or new feet have been put on too high up. No bottom row to the banding is an almost certain sign of an altered base.

If you wish to shorten a long clock, there are two ways in which to approach the problem, from the top and from the bottom. One thing to watch for is the sawing-off of the 'horns' from a Nicholas-type case. The man who did this left a flat top right across from side to side and added a new moulding across this top: and the result was a clock a good 6in or more shorter. Shortening the hood and base can bring a 7ft 8in clock down 12in or more in height. This was often done, to get a tall clock into a low-ceilinged cottage.

There was another answer to the problem and this I have heard of as an occasional practice until quite recent times, by which I mean within the last twenty years or so. I have talked to numerous people who recall this happening in the homes of their parents. The trick was to cut a hole in the floorboards and to sink the base of the clock into this hole to whatever depth was needed to get the top clear of the ceiling. With a stone floor of course you had to remove one flagstone and dig a hole in the soil beneath. Another version of the same trick was to cut a hole in the ceiling to let the 'horns' through and, ridiculous as it may sound, I once did buy a clock that had been in that very position.

Now for the Lomax case (see p 94). One is tempted to think that this would have been made by the famous cabinetmakers, Gillow's of Lancaster. It does have several features which one sees on cases by Gillow and it *could* have come from that firm. There were, however, many other cabinetmakers who could produce work just as excellent as that from the Gillow workshops, provided the customer would

Page 111 (*left*) William Terry, Bedale, 1787; (*centre*) Caleb Nicholas, Birmingham, c 1785; (*right*) Forster, Sheerness, c 1790. A clock with a traditional southern top

Page 112 (*left*) John Palliser, Thirsk, 1817. Note multi-pointed door top and semi-French feet; (*centre*) Palliser, c 1820, with break-arch hood top; (*right*) Robert Gillies, Beith, 1802

pay for the cost of the workmanship and materials. After all, Gillow's were only working to designs that were already fashionable.

Personally I rather object to the attitude whereby a chair, which for instance happens to have features that Chippendale copied, is called a *Chippendale* chair, implying not that it was made by the man, but that it was based on his designs. Such a term also imparts a suggestion of the Chippendale fame to that particular chair. In fact Chippendale was merely translating *existing* tastes and designs of his time into 'published' form and relating them to furniture, and in some instances, as with his designs for tall clock cases, quite impractically. In this instance this Lomax clock case has no need of the Gillow name to enhance its standing; its excellence is there for all to see, regardless of who made it. But perhaps we could say that it is a case of a style which Gillow's *also* made.

It is obviously a beautiful piece of work. It is made of solid mahogany with the front veneered in choice pieces. Its shape is typical of its period for a northern clock with the exception of the hood pediments, which are in a less common design. I have had clocks with this type of pediment, though not many, and mostly they were in the 1770–90 period. The oval opening to the hood door is forced on the cabinetmaker by the shape of the dial, and this oval dial shape is extremely rare.

What a fine case this is, standing exactly 7ft 8in high. It is a clock that was up-to-the-minute for fashion: the newest white dial, probably made in this special shape for an owner who wanted and could afford the best; a case of superb craftsmanship, in the finest woods that money could buy; reeding on the trunk pilasters and hood pillars, which are capped by Corinthian brass capitals; ogee bracket feet; all these would mean extra time and cost in the making. Its date is about 1785–90, about the same time as the next case, that of the John Bancroft clock (see p 94).

The Bancroft case is a country cousin to the Nicholas and Lomax ones. It lacks much of the fancy brassware, having just simple brass caps and bases to plain columns. Its trunk sides are square—no quartered and reeded pilasters here. The square trunk sides are an earlier feature retained by the casemaker for simplicity. The wood is straight-grain oak. The case would originally have had bracket feet, though these are now missing. The door takes up the whole trunk length, the top of the door being arched in imitation of the dial shape. This is a quite common shape for a door top in this period, again carried through from earlier times.

The door and base panel are cross-banded in mahogany. Swan-necked pediments to the hood complete a very pleasantly proportioned and neat case, standing exactly 7ft high. Small decorative features which barely show in the illustration are an oval stringing of light wood in the centre of the base panel and quarter-circle (fan) inlays of light wood in the corners of the base panel. While the Bancroft door top follows the dial in shape, clocks with square dials naturally tend to retain the flat door top for a longer period, and then eventually these too adopt experimental shapes. Square-dial clocks tend to be less common, and especially by about 1820–30 they are dying out.

I am unable to illustrate the cases of the Day and Rowlands clocks.

The Forster case (see p 111) is made entirely of solid mahogany except for the softwood backboard. The design is pleasing though plain. There is nothing fussy or fancy, but the case is in the best wood, which in the solid is virtually immune to woodworm. The pillars are reeded, as are the chamfered trunk corners. The plinth shaped into bracket feet is typical. Shaped back splats to the hood are an interesting variation. This design of hood top, while not rare, is a little unusual and more common in the south than in the north. The total height is only 6ft 8in, which is smaller than normal, making it very acceptable in a present-day home. Notice that there is none of the spectacular veining of the Lomax clock in this wood: the grain is relatively straight. Very ornate grains are usually sliced into veneers and solid mahogany cases are usually made of wood with a less exciting grain.

CHAPTER 9

CASES: PERIOD TWO, 1800-1830

The next case (see p 112) is one in solid mahogany, housing the clock by Robert Gillies of Beith, near Glasgow. A solid mahogany case, by the way, is always much scarcer than one in veneer. There is no debate about the age of this clock, because Gillies wrote his name in ink underneath the seatboard: 'Robt. Gillies, Clock Maker, July 1802'. This is not proof of the case's being original, but it is convincing circumstantial evidence, especially as everything fits as being in period. There are several unusual features about the clock. When I first saw it I thought it had a Scottish look about it. I bought it and, on my showing it to a friend who is a clock enthusiast, he immediately said the same thing. At this time we had not seen the name on the dial, which was very worn away and has been repainted since.

It is Scottish then, but why? I am not sure I can say just why. The Scots casemakers seem to have been keen to do things in their own way, following the fashion of the period, but somehow doing it rather differently. For instance, the hood pillars and quarter pilasters of the trunk have twisted *fluting*—the quarter pilasters of an English clock would tend to have reeding there. The English version would be convex, while the Scots version here is concave. Not an important feature, but a noticeable and unusual difference. I had never seen fluted reeding on a clock before I saw this one.

Notice how the hood pillars are attached—on to projections above the hood door. This is a most unusual method which I have seen used on other Scottish clocks. This feature is none too practical, since it prevents the hood door from opening more than 90°. Then again many hood doors (not having this feature) will not open farther than 90° because of the case design. It is not that one has reason to open a hood door farther than this. It is simply that in opening it carelessly one might cause some damage by the sudden and unexpected contact of surfaces.

The fretted pediments would have been of the swan-necked type, which are not a common feature. It seems that someone may have cut down the tops at some time in the past, and they have been replaced with straight moulded capping. The total height of the case is 7ft 1in.

The bracket feet are replacements, and since the base panel has cross-banding around the top and sides only, this suggests that the base has been shortened, perhaps when the new feet were fitted, though probably not by more than an inch or two. The most unusual thing of all is the proportion, the height of the hood taking up a far greater part of the total height than one would normally expect. Whether this aspect may be said to help the overall design or not is a matter of personal taste.

I show this clock as an example not of the typical, but of the design of the period as modified by the Scottish casemaker exercising his own independent whims and tastes. Don't misunderstand Gillies's casemaker. He knew how to make clock cases well enough, and he did not skimp the work he put into them. He knew how the English made their clocks, which may be the very reason why he did it *his* way. Whether we may like the case or not, we should at least respect his courage. The measure of how far the design succeeds is for you to judge for yourself.

I must mention the frets in the sides of the hood, backed by (new) velvet. These are for the same purpose that we put frets in a radio set: to let out the sound. Here the sound would be that of the bell. Such frets are rather unusual in English longcase clocks of the white dial period, but they occur regularly on Scottish clocks. Occasionally one sees a clock with glass panels in the position where the frets are on this clock. The popular view is that such glass panels are replacements where original frets may have been broken in the past. This may well be so, as glass side panels seem to me to be features which serve no purpose. They do not assist even in setting up a clock, when the hood would be removed anyway. Still, one does occasionally see them and we cannot be sure that some clocks were not made with them originally. Glass panels in the sides of hoods may perhaps have been provided, so that one could see by glancing inside whether the clock needed winding, as this would be evident from the amount of gut left on the winding barrels. It would save unlocking the trunk door to see how far the weights had descended. If my suggestion is correct, it would preclude the existence of glass side panels in the hoods of thirty-hour clocks, as these would not show the need to rewind this type of clock.

The case of the first Palliser clock (see p 112) was made in 1817. It is neat, about 7ft tall, basically an oak case with mahogany bandings and occasional inlays of fancy woods. It has swan-necked pediments ('horns', if you must), a feature very popular right through the white dial period. The case proportions are quite reasonable and well balanced. The pillars are simple, plain rods with plain brass caps and bases—nothing like the Nicholas case. The door is slightly shorter, but still a long door taking up much of the trunk. The proportion of trunk above the door is greater than on the earlier cases illustrated. Notice the pointed door top. Flat door tops are now developing into more interesting shapes, not necessarily more attractive, more exciting, or more difficult to make, but really less sensible, as slender points like these are prone to damage, and often one sees the tips broken off. Notice too the lock and ivory escutcheon as an inlaid diamond shape. This ivory diamond came in about 1810 or so and was amazingly popular. One sees it very often on chests of drawers of Regency and early Victorian period. On clock cases it survives right through to about 1860–70, and is quite common.

A point about locks on clock cases may be worth making. There would seem to be more sense in fitting a lock on the case of an eight-day clock rather than on a thirty-hour one, as one would hardly take the trouble to lock and unlock the door every day when winding.

Yet one does see locks on thirty-hour cases quite often, and frequently, as here, on cases which seem quite original to the clock. It is a possibility that the clockmaker had a steady order with a cabinet-maker to supply cases at regular intervals, maybe one a week, or one a fortnight, according to his needs. It seems that the clock production of an average clockmaker working single-handed was about one clock every two weeks, roughly between twenty and twenty-five clocks a year. It is possible that he may have put his clock into whatever case suited the customer from the stock he had available at the time, and hence would arrive at a thirty-hour clock in a locking (eight-day?) case.

The Palliser case has two front feet only, no back feet. The feet are of an intermediate type, halfway in style between the earlier bracket feet and the later splayed feet, often known as 'French' feet, which we shall see on later clocks.

Notice the chamfered corners on this case, and also on the second Palliser case and the one by Fothergill. These are a compromise, a little fancier than the square (Bancroft) corners, but easier to make than the reeded quarter pilasters of the Nicholas clock. Chamfered corners are quite a common feature in the first half of the nineteenth century.

The case of the first Fothergill clock was probably original but it was in very poor condition when I bought the clock. Having spent some years in a hayloft it was very shabby and badly infested with woodworm. We have no photograph of it but it was a typical case of the period (1805). The interesting point about it was that it had shell inlays. I don't know the reason for it, but shell inlays became very popular on furniture about the turn of the century. There was a fashion lasting about twenty years or so for shells. Shell-like decorations appear in the corner paintings of dials and often on cases, where there is frequently an inlaid shell in the centre of the door, with sometimes another one in the centre of the base. Shell decorations usually appear about 1790–1810, and they continue for some years. An alternative popular theme for these door-panel inlays was a motif of Britannia with a lion, and sometimes a simpler rosette or wheel motif. This Fothergill case was similar in style to the case of the clock by John Farrer of Pontefract, though with reeded pillars.

The second Palliser case is perhaps a slight reversion in style: height about 7ft, flat door top, bracket feet (front feet only, again), plain pillars—largely oak with a mahogany hood door and a band above the trunk, also of mahogany. The trunk door does not quite fill the whole trunk length, but it almost does. Chamfered corners to the trunk are also in mahogany. A very simple clock, it is not quite so well balanced, as the base is a little out of proportion. The interesting point about the case is the hood top, which is of a broken-arch pattern. This was an alternative to the swan-necked pediment and a design which I happen to like, though one sees it less often than the 'horned' top. The implication is that the 'horned' versions were more popular at the time. This, by the way, is another clock which is now in the Netherlands.

The second Fothergill case has progressed further. It is still a pleasant enough clock and reasonably proportioned, standing 7ft 4in —almost a small one by London standards! There are swan-necked pediments, straight pillars, and front feet only again. The case is basically oak inlaid with mahogany banding, and the hood face and the trunk are veneered in mahogany with some rosewood. Notice, however, how the door has shrunk in length, leaving almost a square area beneath it, for what purpose I cannot imagine. This shortening of the door in many cases can often spoil the whole look of a clock, and there is no practical gain or economy in having done this. If anything, a short door makes it harder to get the pendulum in when setting up the clock.

This Fothergill case (about 1825–30) still has the flat door top and chamfered trunk corners retained from a slightly earlier period. It is important always with case styles to remember that each feature is subject to change independently of any other feature, and variations will occur according to the whim of each individual cabinetmaker. A point also to be borne in mind with styles of both cases and dials is that a feature of a past style can be reintroduced by any designer who may wish to revert to that earlier feature. The opposite, however, does not apply, ie a designer will very seldom incorporate a feature that has not yet been tried out and 'accepted'. For example, a casemaker of 1860 could if he wished revert to the plain round pillars as on the second Fothergill case. A casemaker of 1790, however, could not possibly use pillars of the reeded twisted type on the J. C. Elliott case. They would have been utterly out of proportion and, more importantly, they had not yet been designed for use on a clock: no cabinetmaker would dare jeopardise the sales potential of his goods by experimenting so boldly. Experimentation tends to be a slow and gentle process.

Some conservative clockmakers may have stuck to traditional case styles long after others had begun using newer designs. A clock of, say, 1820 could conceivably have been housed originally in a case of a style more appropriate to twenty or thirty years earlier, though in fact this is unlikely to have happened. I am thinking here of different casemakers in the same areas, rather than variations between regions. Again the reverse is not true: an 1820 clock could not have been housed originally in a case of 1850 style. Fortunately when it comes to spotting swapped cases they are often easy to detect, as the case style is usually later than the clock style: almost an impossibility for an original clock-with-case. A case which has been changed with care and knowledge is almost impossible to spot.

The whole question of whether or not one should change cases is very much open to debate. There are times when it can be a desirable improvement—if, for instance, a case of the wrong period can be replaced by one of the right period. It would all rather depend on the ability to discriminate of the person doing the substituting. In view of the fallible human factor, cases are probably best left alone. In instances where the existing case appears to be original to the clock, it seems a great pity to switch it for another, whatever the motive. I know dealers who switch cases regularly. I have done it myself on infrequent occasions when, I tell myself, I could feel justified in doing so. I must also on occasions have bought clocks which were not in their original cases, without my ever knowing this; and so must most dealers, and even more so, private buyers.

As far as possible I am trying to give positive dates for each item

I illustrate, but of course changes were not as sudden or as precise as may sometimes be implied. Obviously, a style did not end in 1820 and a new one start in 1821.

Several points about the John Farrer case, we see next, should strike us as different from those we have looked at previously. First and most obvious, it has no feet. The Bancroft case had no feet—because it had lost them—but almost certainly it was built *with* feet. This Farrer case was probably made with a plinth instead. I have seen many cases very similar to this one, which is a stock type, and they virtually always have a plinth. Why, I don't know. Most northern and midland cases originally had feet. With southern cases one can be less emphatic as they seem quite often to have had a plinth. This Farrer case is one of a type which, while northern, breaks the general rule by having a plinth.

What is so characteristic of this type? Well, this case could have been made anywhere between about 1800 and 1820. It is basically oak, with mahogany banding set in about 1in from the edge of the door and base panel, mahogany veneered trunk face and hood door and surround to the base panel. It has swan-necked pediments and shell motifs. These items categorise it, especially the shells. At this time there is a tremendous vogue for shell inlays, almost always in the door centre and centre of the base panel. The door inlay is usually a shell, sometimes a plaque of Britannia with a lion. The base inlay is often, as here, a 'rose' motif or 'ship's wheel' pattern.

The brass finial with the eagle is probably not original, as they often got broken and lost over the years. It *could* be original, but it is more likely a replacement, though not a modern reproduction one. Brass finials can be bought as new today, in traditional style, usually produced by concerns who make cabinet handles.

One unusual feature of this case is that the hood pillars and quarter-trunk pilasters are inlaid with stringing lines of black wood, a nice little touch making the case just that bit different, though this is not a very important feature. The clock belongs to a friend of mine who is a clock restorer, and while it is in no way outstanding, it is a good example of its kind.

The case of the Roberts clock (see p 130) was made about the same time as the Farrer one, about 1820. It stands exactly 7ft high, a neat example of a north-country clock. In size it is typical of most northern clocks of this period. The very large provincial cases were mostly made after 1830, as we shall see later; but this Roberts case is still reasonably proportioned and small enough to go into any modern home. The size and proportions are typical of northern clocks, but the style is unusual in so far as the hood shape is concerned. This pagoda-top style is more usual on southern cases, though not rare on northern ones.

The case is original. On the back is pasted the original delivery label, but unfortunately it does not tell us the name of the casemaker. It reads:

> Mr. Wolster* Roberts watch and clock maker Derby
> two clock cases to be forwarded
> direct with A Derby Boat
> with great care and kept Dry
> By the Nottingham Steem
> Packet to Gainsbro.

The case is in solid mahogany, the front veneered with selected figurings. Cases of solid mahogany made (in this example indisputably) for white dial clocks—in this instance, too, brought some considerable distance—show that there can have been no feeling at that time of white dial clocks being 'second-class' clocks. Apart from this illustration of the trouble taken over such a clock, at this period if you wanted a longcase clock, this is what you had. Brass dial clocks, while still being produced here and there, were very largely out of date by this time.

The white lines are inlaid *stringing* with paler woods. Notice the double-arrow stringing line on the chamfered trunk corners. These chamfered corners very commonly carry fancy inlay decoration from this time. The ivory keyhole escutcheon is also a typical feature from

* Notice that the label miswrites the name, which on the dial is Woolston Roberts. Or is the error in the spelling on the dial?

now on. The hood pillars are still reeded with brass Corinthian capitals which, however, are not commonly used as late as this. A wavy-topped trunk door is typical. The base tapers off into feet which are now an extension of the base itself, not separate as were the bracket-type feet of a few years earlier. It is in absolute contrast to the next case, though of the same period. Many a case of this Roberts style is now seen housing a late eighteenth-century brass dial clock, because a dealer thought he could make more out of it that way; also collectors wanted good cases, but not if they had white dial clocks in them. A few misinformed writers produced a misinformed market of 'collectors', and to meet their needs dealers destroyed many original white dial clocks by robbing them of their cases. The finer the case, the less the chance of its being allowed to remain with its white dial clock.

The case of the J. Keys clock (see p 130) is undoubtedly original. Inside the door is *branded* 'J. Keys' and written in pen alongside this is 'April 15 1825'. The same wording is on the backboard inside. On the inside of the movement frontplate is scratched 'Made by Joseph Keys April 15th 1825' (also: 'cleand February 15th 1832' and 'cleand by ditto August 13th 1834'). Maybe Keys was farsighted enough to do this on purpose, to prevent the case of his clock from being swapped.

It could be argued that this is not proof. However, when taken with the details of the clock and its maker, it is consistent. The question of faking hardly arises, as no one could possibly have gained financially by putting a better case on such a clock than the one it was made with. Any tampering with or faking is almost always done for financial gain.

The case is made entirely of solid mahogany, except for the backboard which is oak. The plinth is a replacement: probably the case originally had a bracket-foot type of plinth, ie a plinth shaped into a foot at each side. It stands exactly 6ft 3½in high. It is a very neat and simple case: nothing fancy. Two things are remarkable about it. To begin with, it is a simple thirty-hour clock, yet obviously in the opinion of the purchaser or maker it merited a solid mahogany case.

Not that the case itself is very special. However, it would have been cheaper to make one in oak or deal. The fact that this clock was thought at the time to deserve mahogany may indicate that a thirty-hour clock was perhaps not looked upon as the poor relation to the eight-day one in the way that is sometimes suggested. Evidently it was considered more important to spend extra on a mahogany case, rather than have an eight-day clock in an oak case. All I am saying is that the standards or values by which we judge today are perhaps not the same standards as those of the time. It is often true of a country clockmaker that his normal production was a thirty-hour clock and that an eight-day one was the exception, not the rule. This is not the place to go into a detailed argument of that point. However, I have a record of over 100 clocks made by an above-average late eighteenth-century clockmaker and of these only fifteen are eight-day ones.

Records I have of another clockmaker's output at the same time show that, out of a total of almost 300 clocks he made over a twenty-year period, only six were eight-day ones. Six eight-day clocks in twenty years is an extreme example, but I mention it simply to clarify that for most people a clock meant a thirty-hour clock. Those who were prepared to pay twice the price to get an eight-day one were in the minority in the eighteenth century. In the nineteenth century, however, the trend changed and increasingly more customers preferred to pay for the costlier eight-day rather than accept the daily duty of winding a thirty-hour one.

The other point about this case is its style, which has barely changed from that of fifty years earlier. Notice the long door, the three-quarter pillars attached to the hood door, and the quarter pillars at the back of the hood. That is really all the style there is to it. The door, you will notice, has the best piece of mahogany with the finer figuring. The sides and base are plainer straight-grain. As we shall see shortly, the case of the clock by another Devonshire maker (of Dawlish) is also in this similarly archaic style. These examples teach us, if anything, how hard it is to draw firm conclusions about case styles. There is no reason why a particular maker should not have kept to the styles he personally liked, and of course a clockmaker aged sixty could still retain a fondness for, or even be hidebound by, the styles of the days of his apprenticeship. The same thing can happen in a whole district. A quiet country area like Devonshire need in no way feel the need to imitate up-to-the-minute fashions prevalent in industrial centres such as the midlands or north of England.

It takes a lot of practice to be able to say with any certainty that a case is right or wrong just by looking at it.

One point I have not mentioned could be argued. Suppose the case were *older* than the clock, and that J. Keys sold the clock in a second-hand case? That could explain the old-fashioned style, admittedly. However, one must remember that Keys was selling a *new* clock, not an antique. Could he have got away with an older case? I doubt it. And then when we think that Keys was proud enough of the work to put his name on the movement, the dial, and twice on the case . . . That was a new case all right, and it is original.

Unfortunately the archaic nature of the case style which Keys used was its downfall. I sold this clock some time ago. Recently I was surprised to run into the case again in an antique shop, but now containing a thirty-hour *brass* dial clock by Abraham Fell of Ulverston, made about 1760–70. I could recognise it as unquestionably the same case because of various minor repairs to it, which I had personally carried out. On account of the older-than-period style of case it is unlikely that anyone will now know that it was made about sixty years later than the clock it now houses—unless they read this and can spot the marks where Keys had once written his name inside the case. Incidentally, the price of the clock had multiplied almost by three in a period of less than a year because of the switched movement and dial.

Since a conservative clockmaker, if he so wishes, can retain features of earlier traditional styles, we can find other examples similar to this Keys case, and we grow to accept the possibility of what is termed a 'throwback' to earlier styles. What no clockmaker could do, of course, is to use case styles which have not yet been conceived. While we might find an 1825 clock in a 1795 type of case we could hardly find a 1775 clock in an 1825 type of case, that is in areas where such styles have changed during this period.

In attempting to assess the originality of a case, it is always as well, if you begin to doubt, to look for the motive that might have led to a case being changed. Remember, the older the case, the harder it is for a dealer to come by a spare one to fit a caseless movement. Sometimes he will have to make do with one less old than he would like. Hence if a clock seems to have a case *newer* in style than one might expect, you might sensibly regard it with caution. If a clock seems to have a case *older* than one might expect, here a little more tolerance is needed, because of the possible *throwback* of case styles we have been discussing. In thinking along these lines we have to credit the dealer with the ability to recognise cases of varying ages, and there will be occasions where we may be giving credit where it is not due. Anyway, I still feel that it is as well always to bear in mind the motive for a change you think might have taken place. Has there been a financial gain to the dealer by such a change, either now or in the past?

Recently at a very highly esteemed Antiques Fair, I saw a provincial clock of about 1730, a brass dial one of course, in a solid mahogany case of quite obviously late eighteenth-century style. The description ticket gave details of the maker's known dates. A solid mahogany case as early as this would be rather exceptional anyway, as the wood was not imported in any great quantity until 1750 or so. However, the case was typically 1780 or 1790. The price was several hundred pounds. It is as well to look carefully, not just appreciatively.

Also illustrated (see p 129) is the case of a clock by Peter Penman of Dunfermline, Scotland, made about 1820–30. The dial is circular, plain, with Roman numerals only, matching brass hands of crown pattern, and no falseplate. The case, while at first sight of an unusual shape, is a style met with occasionally on Scottish clocks. The pointed hood top lacks finials above each pillar; it did have them once as the holes are still there. The door is a circular brass one, an attractive though impractical feature, since these heavy brass doors wear out their hinges by their own weight. The trunk door is convex, a frequent feature of Scottish casework. The carved decorative piece, which looks like the door top, in fact is attached to the trunk, and the door top is flat. This is a sensible idea as over-elaborate door tops are very prone to damage. This point is worth mentioning again, as it is apparent that casemakers eventually began to recognise this factor themselves, as in this example. Full-length trunk pillars are an interesting feature of this fine case, which stands exactly 7ft 2in to the top of the centre finial. The case has solid mahogany sides and superbly figured mahogany veneers to the front. Scottish cases are often of high quality in both woods and construction.

CHAPTER 10

CASES: PERIOD THREE, 1830-1870

The first case we examine (see p 147) is that of the Stephens & Davies of Neath (Glamorgan) clock, and a very pleasant case it is too for this period. It is made of solid mahogany with veneers on the front. Not too large, it stands about 7ft 6in high. The proportions are good, well balanced, traditional. There is still a long door, which by now is an unusual feature except with conservative casemakers who, as here, retain the earlier type. Notice that the hood door has scalloped (wavy) edges, this also being a feature more usual on clocks of a slightly earlier period. The pillars are turned, generally a late sign, and the splats at the back of the hood are shaped, which latter feature can occur at any period on white dial cases. The swan-neck scrolls have very exaggerated curls, though I find these attractive. The hood pillars have brass Corinthian capitals—rather a late use of these, for one normally associates them with the late eighteenth century. When compared with the cases by the Snows, of almost exactly the same period, this one certainly has the better lines and proportions. I show it as an interesting example, not as typical of most Welsh cases of this time which would perhaps tend more towards the style of the Snow cases.

The case of the Dawlish clock (see p 147) is most interesting because of its apparently early style. It stands 6ft 6in high, and is mainly of straight-grain oak with mahogany-veneered hood door, mahogany bandings to the trunk door, and the same to the base panel. Pillars are still, even at this late date, attached to the hood

Page 129 (*left*) John Farrer, Pontefract, c 1820. Note wavy top to door; (*centre*) John Fothergill, Knaresborough, c 1825. Note chamfered trunk corners; (*right*) Peter Penman, Dunfermline, c 1820–30

Page 130 (*left*) Woolston Roberts, Derby, c 1820, with bell-shaped hood; (*centre*) Joseph Keys, Exeter, 1825; (*right*) Thomas Snow, Knaresborough, c 1835

door, where one would normally expect free-standing pillars by now. Bracket feet are also carried on here into a considerably later period than one would expect. The brass escutcheon is a modern replacement. I bought this clock in a very dirty condition, and it could not have been working for many years. In view of this and also in view of the unusually small dial aperture of 10½in square, it seems unlikely that it was a replacement case. I cannot be certain that the case is original to the clock, but it seems very likely. It is interesting to find, as with the Keys clock, a case of such early traditional style being made as late as this: 1835 or even later. Such a case one might expect in the 1780–1810 period, but 1835 is unusually late for this style. The dial also had several features which were reversions to a style of twenty or more years earlier.

The Hooker of Lewes clock we cannot show a case for. It had no case when I bought it, nor when I sold it. However, I did see a photograph of the case of the second, arched-dial clock I have mentioned by that maker, and its style too was vastly earlier than one might expect for 1835 or so, being very similar to the case of the clock by Forster of Sheerness, Kent (c 1790)—in other words, unchanged in almost fifty years.

The next two cases are by the Snow brothers, Thomas of Knaresborough and Richard of Pateley Bridge (Yorkshire). In many ways they are typical of north-country cases of this period, 1835–40. They are relatively tall, measuring about 7ft 6in. A clock of this type standing much over 8ft high is quite exceptional and usually confined to very late cases of after 1850. They have high bases and therefore shorter trunks with short doors set into these trunks, so that quite a wide expanse of case surrounds the door. This is the type of high-based, short-door case which, while typical of Yorkshire clocks after about 1820–30, is frequently mistakenly cited in reference books as being a normal feature of Yorkshire clocks from about 1770 onwards. Clocks with this type of short door, as in the two Snow cases, were made in Yorkshire and most other counties too of the north and midlands from about 1820 or 1830. Earlier than that they would be extremely unusual.

These cases have swan-neck pediments, ivory keyplates, 'French' feet (at the front only)—the Knaresborough case feet have been shortened—turned hood pillars and wavy-topped doors. All these points are typical of *this* period. The Knaresborough case is in solid mahogany with a veneered front in handsomely veined pieces, and it carries much inlaid stringing of lighter woods. It has flat canted corners to the trunk, which flats are inlaid with stringing lines, and it has criss-cross stringing above and below these canted corners.

The Pateley Bridge case is in oak with mahogany and rosewood veneers on much of the front, but is still largely oak. This is more a country case, but by this time very much imitating the style of a town case. The pillar turnings are heavier, and the general build is slightly heavier. This is the case of a thirty-hour clock and Richard Snow could and did frequently supply better cases with his eight-day clocks. The latter tended to be all-mahogany versions of his thirty-hour cases; better in materials, but of the same overall shape and style, which was *the* fashionable style. Thomas's cases were generally a little more graceful, perhaps that bit sleeker and more slender. Richard's case also has twist-turned quarter pilasters to the trunk, a feature which crops up quite often from this time on: one sees it also on chests of drawers in this area. The oak of this case has a yellowish look and contrasts strongly with the mahogany. Early oak cases tend to be darker in colour, the original (usually stained) oak having darkened further after years of having been polished with wax, and through dirt being polished in rather than cleaned off. Cases of the Snow period, however, were often treated when new with a thin French polish type of finish, and this tends to preserve the original colours by preventing wax from seeping into the wood. The strongly contrasting yellow and dark brown of these cases must have been popular at the time, though personally I cannot say that I care for it.

As a grand finale we come to the case of J. C. Elliott's clock (see p 148). Elliott worked at Blyth in Northumberland early in his career, I discovered recently, and moved to Leeds probably not long before 1850, when his shop was in Leighton Lane. This clock was made about 1855. It stands exactly 8ft high, yet looks very much bigger than this. Some people would find this clock repulsively ugly—overfussy, tasteless, disproportioned, yet fascinating. When I first saw this clock in a gloomy attic, it very nearly frightened me to death, but I bought it. Carrying it single-handed down three flights of stairs did little to help me recover from the shock.

A few people might find this case beautiful, though they would be in the minority. While not exactly typical in every detail, it does represent a certain type of late case to be found in the north. When it was made it was one of the best examples of its kind—whether or not we happen to like this type of clock today. Many hours of the most skilled work went into its making, as did the choicest woods available. It is solid mahogany with the front veneered with rosewood, mahogany, and with marquetry inlay in lighter woods. The feet are very shortened, stumpy feet of the 'French' type: front feet only, again. Notice the enormous reeded, twisted pillars to the hood and similar though larger ones in the trunk corners. Full pillars on the trunk and even on the base corners were very fashionable at this time. Look at the very elaborate shape of the door top.

The biggest fault really is the proportion, the trunk now being much shorter in relation to the hood and base. The clock is too wide for its height; at least, this is what we may feel today. It is interesting to wonder what opinion might be in 100 years' time. This clock, like so many others, is now in the Netherlands.

The men who made this case are long dead, and with them died craftsmanship of the highest quality. The design of this case may be criticised, but the execution, the workmanship, are such that we shall probably never see the like of again. Anything involving the materials, skill and time that this case involved would today be thought far too expensive to be undertaken. How many cabinet-makers are there today who could make a case like this, and at what price?

In the photograph (see p 148) this Elliott case is shown alongside one of c 1800–10 containing a clock by Richard Morland of Kirkby Malzeard, Yorkshire. It is useful to contrast the two, especially in regard to the proportions. The Morland case was of oak with mahogany bandings, a shell inlay in the door and reeded quarter-trunk columns of mahogany. The real difference is not so much in the extra height of the Elliott case, but mostly in its extra width and its much higher base with consequently shortened trunk length and squat door. The Elliott case must have been far more costly to make, but today most of us would prefer the Morland case.

Cases of the Elliott type are exceptional and not typical. However, if we take the essential features of the Stephens & Davies and Snow cases, these set the pattern for most casework of the period. Long-case clocks after 1840 are less common. In the south they are relatively infrequent; in the midlands and Wales they persist; the stronghold after 1840, however, is mainly in the north with Yorkshire and Lancashire by far the most prolific areas at this time.

This last fling at lavish casework was perhaps the only answer that British clockmakers had in the face of the competition which now threatened them and finally crushed them in the form of imported clocks. The American shelf and wall clocks, the imported 'Vienna' wall clocks, were pouring into the country. Against these 'miniature' mass-produced clocks, the British product could not hope to compete either in size or price. Perhaps the only hope was deliberately not to try to compete, but to go to the other extreme in making the longcase more lavish than ever. Cases like the Elliott one illustrate how far the clockmakers went in this attempt. But the attempt failed.

I have bought these large cases of the Elliott type on occasions purely to break them up for the veneers, and tragic as this may seem, it is true just the same that only a very few years ago there was no market at all for clocks of this extreme width. No doubt we shall be criticised in the future for this, just as we now criticise those who weighed in brass dial longcase clocks for their value as scrap brass fifty years ago.

> My grandfather's clock was too tall for the shelf,
> So it stood ninety years on the floor.

The well-known song by the American, Henry Clay Work, was first published in 1876 (E. L. Edwardes records in his excellent book, *The*

Grandfather Clock). Perhaps the first line of the song sums up the problem. Ironically, the song that gave these clocks their lasting name also sounded their death-knell.

Space will not allow illustration and detailed description of every variation in case design, but the principal aspects have been covered. The following reference table attempts to summarise in brief form the major aspects involved in the dating of case design.

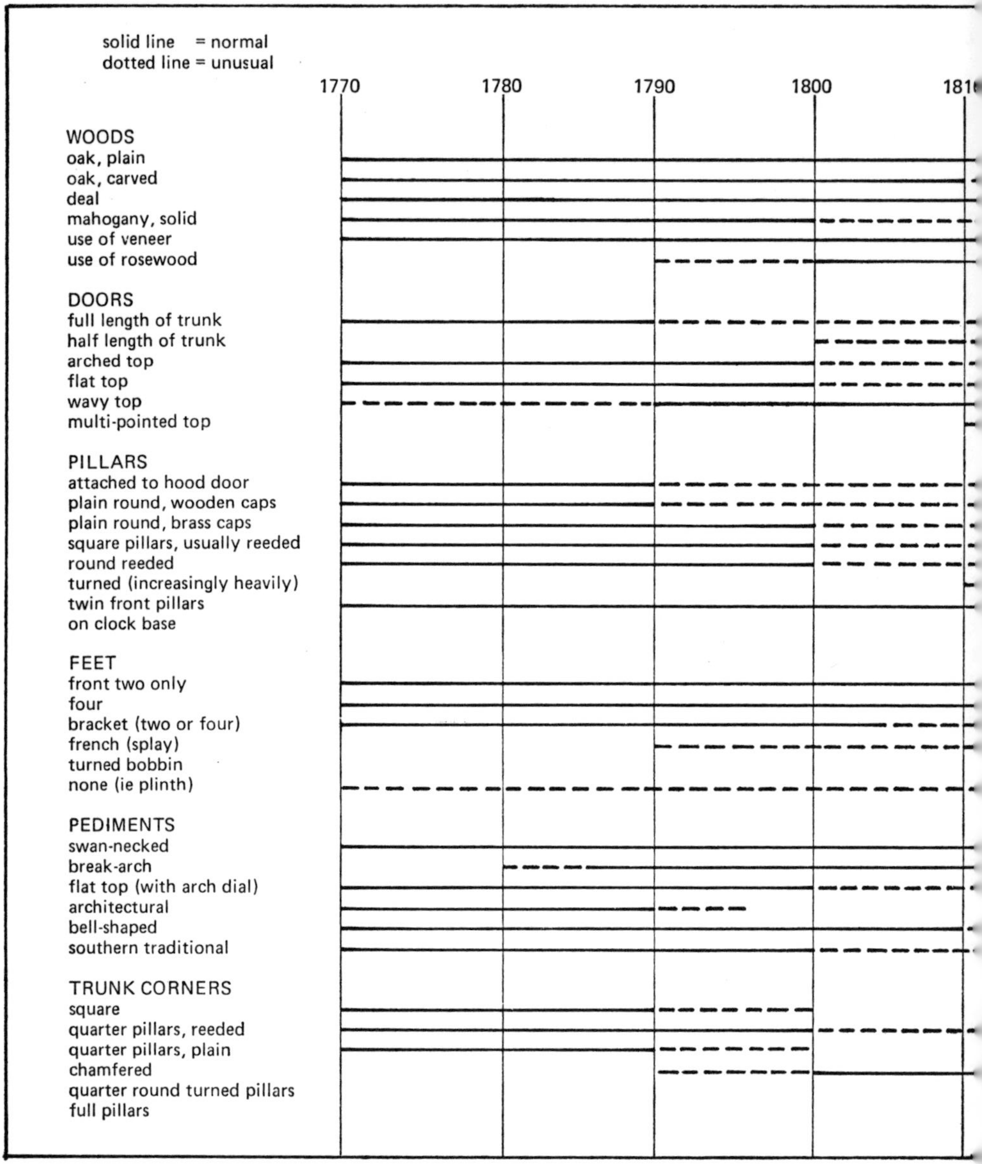
solid line = normal
dotted line = unusual
1770
1780
1790
1800
WOODS
oak, plain
oak, carved
deal
mahogany, solid
use of veneer
use of rosewood
DOORS
full length of trunk
half length of trunk
arched top
flat top
wavy top
multi-pointed top
PILLARS
attached to hood door
plain round, wooden caps
plain round, brass caps
square pillars, usually reeded
round reeded
turned (increasingly heavily)
twin front pillars
on clock base
FEET
front two only
four
bracket (two or four)
french (splay)
turned bobbin
none (ie plinth)
PEDIMENTS
swan-necked
break-arch
flat top (with arch dial)
architectural
bell-shaped
southern traditional
TRUNK CORNERS
square
quarter pillars, reeded
quarter pillars, plain
chamfered
quarter round turned pillars
full pillars

LOCK CASE FEATURES

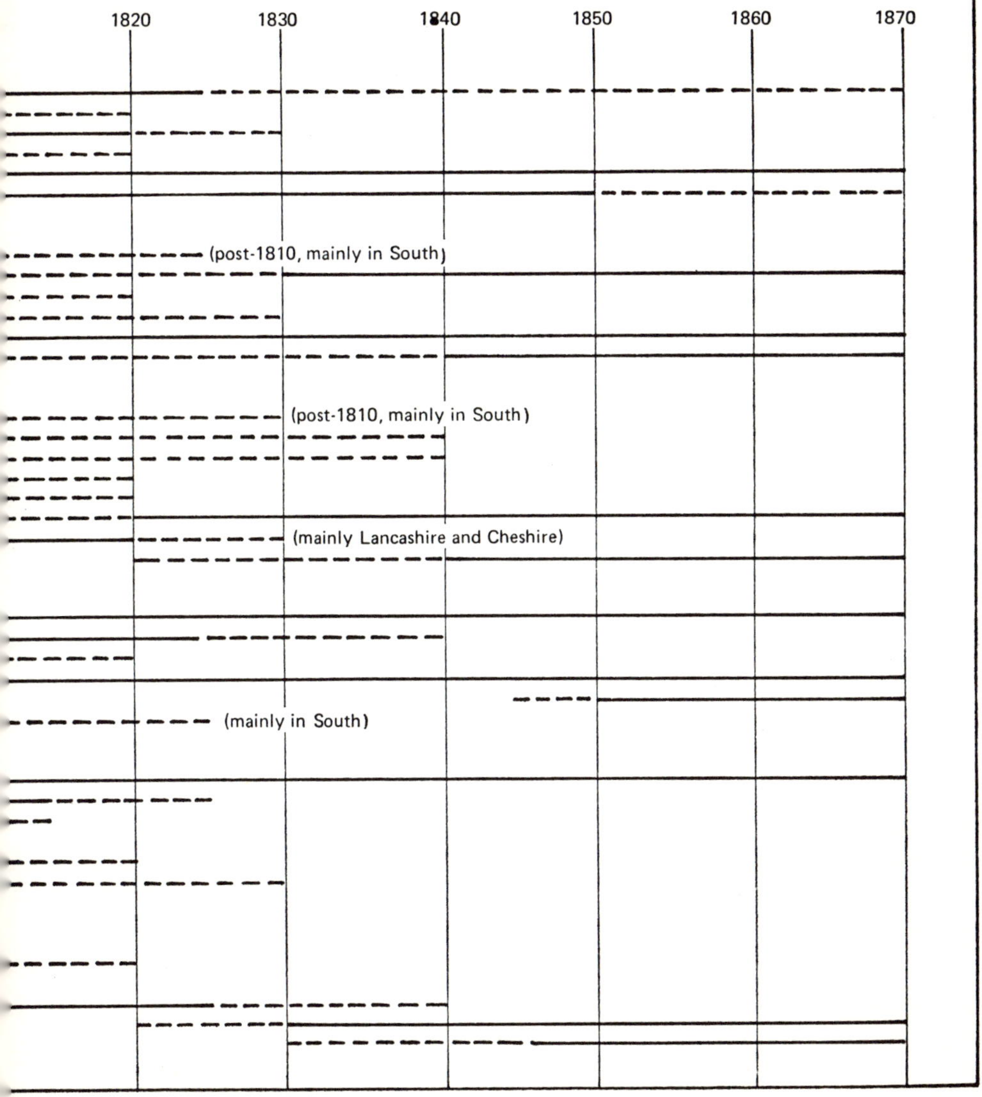

CHAPTER 11

PENDULUMS

A pendulum is not really an item by which one can positively date a clock, but it can help, and so a few points might be worth mentioning. The pendulum in any eight-day longcase white dial clock you are likely to see will be a 'one-second' pendulum, which means that it beats every second. Its length will be a little over 3ft 3in, which is as technical as I intend to be, and which will serve to enable you to recognise it when you see it. With thirty-hour clocks there is more variation in the length of pendulum according to the gearing of the wheel trains, and anything between 2ft 5in and 3ft 10in in length would be possible, or even wider variations. Any white dial clock you see with a pendulum considerably different in length from this will be something very unusual and rather than measure it you would be well advised to call in an expert to look at it for you.

On early white dial clocks, up to about 1800, the pendulum is usually made of a length of wire rod terminating with a 'bob' made of lead with a brass (uncommonly, copper) covering on its face side. From the beginning of the nineteenth century the lead bob is gradually replaced by an iron one, its face side painted black, or sometimes faced with a tin covering on the face side and this tin cover painted black. About this time the pendulum wire rod is often replaced by a flat steel rod, ½in wide or so, which form persists right to the end of the white dial period.

After about 1815 or 1820 it is not unusual to find a wooden pendulum rod instead of a steel one (the J. C. Elliott clock illustrated—see p 148—had a wooden rod). These occasionally have, in addition to the normal adjusting nut below the bob, a micrometer adjuster fitted at the point where the bob and rod meet: a simple device designed to raise or lower the bob by a very small amount and more precisely than a turn of the rating nut would allow. The reason for the wooden rod-and-micrometer fitting was of course to improve the precision with which the clock's timekeeping could be set. The wooden rod was less liable to contract or expand with changes in temperature. The micrometer fitting is uncommon but by no means rare; nor is it an indication of anything very remarkable about the clock, but rather an interesting variation. The wooden pendulum rod is quite common.

After about 1820 or so one sees not infrequently a pendulum with the bob-face painted in bright coloured patterns (japanned in fact). There seems to be no special reason for this except to make the bob more colourful in keeping with the dial painting. Such a bob could well have been japanned by the dialmaking concern, and might perhaps be an indication that the dial, movement and pendulum were all bought in direct from the factory by the clockseller. The view is sometimes expressed that such painted pendulums were originally made for lacquer-cased clocks. However, one sees them too often on white dial clocks to accept them all as being replacements. Incidentally, I have *never* seen a white dial clock in a lacquer case, and would feel very suspicious about it if I did.

Another interesting treatment of the pendulum bob-face in this period (post-1820) is the type with a brass-faced bob, the face made from stamped, patterned brass: very ornate and very 'Victorian'-looking, and of course almost impossible to keep clean because of the deep crevices in the stamping.

An interesting type of pendulum, which I see frequently but which seems not to be common everywhere, is one where the bob-piece simply hooks on to the end of the wire rod, which end is formed into a hook. This incredibly simple version is an alternative to the normal attachment where the wire-rod end screws into the bob-stem as a rigid attachment. It is surprising that this idea did not become more widespread, as it very much simplifies the process of setting up a clock by allowing one to slip the pendulum rod into place, and hook on the bob and its stem last of all. This is especially helpful with later northern clocks where shorter door-height can cause the normal pendulum to be bent during insertion. This hook-on bob-piece is usually of the tin-fronted post-1820 type but I have seen it occasionally on brass-faced pendulums and even used on the apparently original pendulums of brass dial clocks as early as 1760.

Pendulums, of course, are frequently *not* original to the clock. It is therefore difficult to determine precise style changes, but these I have given will apply generally. Being a relatively fragile thing, a pendulum is subject to damage and replacement over the years, and no great harm is done if a clock has a replacement one. It is naturally more satisfying if a replacement of the correct type is fitted.

I know one dealer who keeps all his pendulums in a tea-chest and offers his customers the pick of the bunch—steel or wooden, straight or curly, with or without the bob! There is also the true story of one 'horologist' dealer who, after long deliberating the merits of a very ordinary wire pendulum rod on one of my clocks, pronounced that it was made from an old bicycle wheel-spoke. I was quite unable to refute this opinion, and was obliged to admit that I was not well versed in the history of bicycle wheel-spokes, especially of a type made for a wheel which must have had a diameter of 80in or more! However, any reader who happens to have a stock of old 80in-diameter bicycle wheels will be delighted to know that he could find a ready market for their spokes among clock repairers, as their quality is such that I know of only one 'horologist' who can distinguish them from genuine antique pendulum rods!

CHAPTER 12

THE WHITE DIAL IN AMERICA

The development of the white dial clock in America demands considerably more research than I can undertake from this side of the Atlantic. British writers on horology have been accused in the past—not without some justification—of having scarcely glanced at American clockmaking. Our only defence can be that while British clocks were and still are exported wholesale, the reverse is not true, and consequently we seldom have the opportunity to examine American clocks in detail. I am speaking here of grandfather clocks only. In America these are sometimes known as 'tall clocks'.

Very many British-made grandfather clocks were taken to America among the household possessions of emigrant families, the majority of them being of the white dial type. Far greater numbers have been exported by antique dealers in more recent times, and vast numbers of these British clocks still are shipped to America every year. Americans interested in white dial clocks in their own country must therefore pay considerable attention to the development of British-made examples, which may constitute a high percentage of those they will meet with. In the absence of knowledge about the maker whose name appears on the dial, the American who is not experienced in clocks might well be unable to decide whether the clock was made in Britain or America.

When we come to white dial clocks sold by American clockmakers, we must remember that some of the clockmakers were trained in this country, and went out to America fully experienced, with a knowledge of trade suppliers in Britain of whatever parts they might wish to purchase. Such trade contacts in Britain would be most useful should there be any problems in obtaining the required materials in America, as at times there undoubtedly were. The practice of some British clockmakers buying in finished or part-finished goods from others in the trade was already well established long before the white dial came on the scene about 1770.

What all this adds up to in the early period of the white dial clock in America, is that there may not always be an easy solution to the problem of which clocks or parts of clocks were made in which country. To add to this difficulty, white dials were an English innovation and Osborne & Wilson, and no doubt other dialmakers, shipped out to America ready-painted dials on to which the local clockmaker might build his own movements or fit British bought-in movements. The falseplate sold with many such dials was designed specifically for just such situations. P. R. Hoopes in his article in *Antiques*, 1931, suggests that white dials were not manufactured in America until after 1800; and he quotes interesting evidence of white dials 'newly arrived from England' being sold by local wholesalers in 1784 in Baltimore, and in 1785 in Philadelphia.

I understand that American-made white dial clocks (with English dials) quite frequently bear no name of the 'clockmaker'. This would seem to be consistent with the English manufacture of these dials, if ordered in quantity by an American wholesaler for individual sale to local clockmakers. At the time the order was placed, the name of the retailer may well have been unknown and the dials would therefore be left nameless. In England with less distance and consequently less delay in order-processing, the individual retailer could order his dials direct and specify the lettering to be painted on them.

On the subject of falseplates I am advised that these are found on apparently original American clocks with *brass* dials. I am not in a position to dispute this. However, if it is correct that American makers used falseplates in this way, it does not invalidate my view that falseplates were initially made specifically for sale with, and for use with, white dial clocks.

By about 1800 the practice had arisen in America of making movements of tall clocks, many of the thirty-hour type, of *wood*. British longcase clocks with wooden gearing would be extremely rare, with the exception of imported German clocks with painted wooden dials and wooden wheels which one sees occasionally bearing an English clockseller's name. These are mostly post-1830 and of course they are not strictly British clocks. The making of American wooden movements soon grew into a mass-producing business with interchangeable standardised parts. Wooden dials were used too, some early attempts having a printed paper face pasted on to the wooden dial proper. Later wooden dials had the decoration and numbering painted directly on to the dial boards. Clocks with wooden wheels had long been known in Holland and Germany and clockmakers from these countries emigrating to America may have been the source of inspiration to use locally available materials.

The name of Eli Terry is normally associated with the beginnings of mass production of wooden clocks in America. The brass-wheeled clock movement was now more a product for the wealthier few; the cheaper wooden one, usually a thirty-hour clock, was for the majority. It was Terry too who first developed the shelf clock about 1815. Shelf and wall clocks were ultimately and rapidly to dominate the scene both in America and in Britain. American shelf and wall clocks poured into Britain in the late 1840s, rapidly putting an end to the manufacture here of longcase clocks. This may be one reason why this type of clock tends not to be very highly regarded in Britain. Many dealers in Britain today are only too happy to ship them back again.

Casework on American tall clocks begins, as one might imagine, as an attempt to follow the English equivalents. However, factors such as the sheer size of the country, the relative sparseness of the population and the consequently greater distances between clockmaker and customer caused a more rapid, varied and individualistic development of casework in America. This was further varied by the different woods used.

It was not uncommon for tall-clock movements to be sold caseless with the option open to the customer to hang the clock on a wall bracket as a 'wag-on-the-wall' clock. Alternatively, he could order a case for himself from a local cabinetmaker or joiner. Or he might make a case for the clock himself. The result was that some superb traditional cases exist alongside some extremely crude softwood ones, which may present a very strange appearance to an eye accustomed to British casework. While the swan-necked pediment top of the British type was quite popular, the plain arched top of many American clocks was capped with a fretted or crested decorative flourish, the latter sometimes called 'whales'-tails'.

By 1830 the tall clock in America had largely been superseded, as it later was in Britain, by more compact, cheaper, mass-produced wall and shelf clocks. This chapter can do little more than indicate some of the important differences between white dial longcase clocks in Britain and America, yet these differences may help distinguish those clocks imported from Britain from those made in America. While painted dials would be of British make up to about 1790, after that date they could be of either British or American manufacture. The lists of known British dialmakers given in Chapter 3 may help decide which, until such time as lists of American dialmakers are published. A wooden dial, painted direct or with paper facing, would *not* be of British origin, nor would a wooden movement.

CHAPTER 13

BUYING AND SELLING

This chapter will not begin to tell you about the price of clocks. That is something you can only find out for yourself by looking and learning. But it may help you to avoid paying far more than the clock is worth. To a certain extent one can expect constantly rising prices to be inevitable, as the supply of antiques on the market gradually shrinks in the face of ever increasing demand. When one buys, the price may seem high; but as the years pass, a good clock can prove to have been a sound investment. On the other hand a clock which has been tampered with, or incorrectly restored, or even plainly faked, will not prove to be the same sound investment. The beginner's problem is to distinguish one from the other.

My aim will be to give you a few hints about *how not* to buy a clock. First of all believe nobody but yourself. There are, of course, dishonest dealers; but there are also those who may mislead a customer quite unintentionally, out of ignorance. As an example of what I mean I will tell you of an experience I had at the time I was doing research into Yorkshire clockmaking. I happened to go into a general antique shop one day in a Yorkshire market town, and there was a brass-faced longcase clock signed with the name only, no town. Suppose the name on the dial was Obadiah Wilson—no town given. It was quite a pleasant country clock, late eighteenth century, but nothing very exceptional. I mentioned casually that it was a nice-looking little clock, or something to that effect. 'Oh, yes,' said the dealer, 'it's a London clock, you know.'

I was speechless. As it happened it was only a few days earlier that I had been looking up some facts about the life of this particular maker who had worked not forty miles from where the clock was then standing. I happened to know about that particular maker, but even if I had never heard of him, everything about that clock indicated that it was a typical country clock, and I was certain that it had never been closer to London than where it stood at that moment. I could not resist taking this a little further.

'Well,' I said, 'there was an Obadiah Wilson who worked at Trumpton on the Wolds. What makes you think this is by a London maker?'

The dealer was quite unruffled. 'Oh, it is,' he said. 'He is listed in Baillie's book, you know.' Now I did not doubt for a moment that a man with the same (unusual) name was listed in Baillie, as I believe this dealer was sincere enough. But the London maker of that name most certainly was not the man who made this clock.

It is worth noting, by the way, that a clock signed with the name only, and no town, is very often the work of a maker in a small village, the name of which would not have been recognised even twenty miles away. In any case a maker who was paying the high premium of working in London would be proud of that fact, and would not for one moment forget to make the very best advertising use of it by putting 'London' on his clock dials. Furthermore, he was obliged by law to specify the place as London.

'Well,' I said, 'I would have thought that this was the Obadiah Wilson of Trumpton.' Then followed protests that I was mistaken, that it *was* a London clock, that he knew it was a London clock, that he had bought it as a London clock, and 'Well, it just *is*!' There was no point in pursuing this conversation, and I had no intention of buying that particular clock. The real point of the story was that this person was a dealer of some years' standing, with a reputation for honesty. His sin was ignorance, not dishonesty, and I have no doubt that he genuinely believed that it was a London clock because, 'Well, it just *is*!'

Now this was a year or two ago and somewhere there is a proud

Page 147 (*left*) Richard Snow, Pateley Bridge, c 1840. Note twist-turned trunk pillars and 'French' (splay) feet; (*centre*) Stephens & Davies, Neath, c 1835; (*right*) unknown Dawlish maker, c 1835

Page 148 (*left*) John Catcheside Elliott, Leeds, c 1855. Note full trunk pillars; (*right*) Richard Morland, Kirkby Malzeard, c 1800

owner of a 'London-made' clock by Obadiah Wilson. He cherishes his clock, lives in absolute ignorance of its true origin, and praises the name of the dealer from whom he bought it so reasonably. One day he, or his heirs, will try to sell that clock to a specialist clock dealer, and then will be sadly disillusioned, or else will disbelieve that specialist dealer and sell it to yet another dealer who also is as uninformed as the person in whose shop I saw it.

The conclusion to be drawn from this story is to know your subject for yourself and to buy a clock entirely on your own judgement.

When you look at a clock, ask whether it works properly, and listen carefully to the answer. A frequent reply is: 'Oh, yes, I think it goes all right. At least it was going in the house when I bought it from the lady who had had it for many years, so it should be all right.' With experience you will learn that this means that the clock is a duff one.

If you look at enough clocks you will soon be surprised as to what a large number of them once belonged to elderly ladies who cared for them for many years. Eventually you realise that the story is similar to the one about the second-hand car with incredibly low mileage which once belonged to the elderly vicar's wife and never had been driven at more than forty miles an hour.

The answer you ought to be hoping for is that the clock is guaranteed to be in perfect working order. There are only two possibilities with a clock: either it goes or it doesn't. If the dealer's answer is not an emphatic 'Yes', it can only be because he means 'No'. This is what you must assume until it is proved to you to be otherwise.

There are many sources for buying a clock: from junk shops or sale-rooms, from antique dealers or specialist clock dealers, or from private collectors who are disposing of an odd one or two. You could buy from any of these, but you ought to understand first which source would be best for you as an individual, and this will depend on your knowledge of the subject.

Forget all about 'picking up bargains' at ridiculously low prices. There are no bargains any more, as far as you are concerned.

Thousands of dealers spend their whole lives running from pillar to post looking for bargains, and they have far greater experience in recognising a bargain when they see one than has the average private individual. If you come across a clock in a dirty condition in a little junk shop, that is because that is the best place for it.

If you go to a general antique dealer, the clock you get will depend on the class of shop you go to. If you go to a better-class dealer who will guarantee the clock, or to a specialist dealer who will do the same, you will be paying the going price. A dealer sells a clock for what (he thinks) it is worth, or for what he can get, not for what he can get less 10 or 20 per cent. The only way you might get a clock at less than its true market value is if your knowledge exceeds that of the man selling it and you can recognise something special about the clock that the vendor does not appreciate. However, if you pay the going price for a clock from a specialist dealer, you are less likely to be cheated than if you are picking it up for a song in a junk shop.

The same thing applies to sale-rooms. In the top auction houses you can expect high standards to be set, but if you wish to buy a clock at a country-house sale, for instance, you must have the knowledge first to know what you are buying. Otherwise you may get the clock home and find too late that it can never be put into proper working order.

Beware of the kind of advertisement one sees quite often in the 'Items for Sale' columns of local newspapers: 'Private collector has grandfather clock for sale. Phone . . . after six. No dealers'. Sometimes the advertiser is anonymous; sometimes he is the well-known 'private collector'. Not very often does he give an address. The biggest give-away is the term 'No dealers', an expression calculated to excite the private buyer into thinking that he is getting at a buy which is not being made available to dealers.

The real reason why the advertiser puts 'No dealers' is very often because he himself is a dealer, and would be instantly recognised by any other dealers who might otherwise call. Alternatively or additionally, the reason may be that the vendor is asking a price which he knows no sensible dealer would pay, and to save himself the embarrassment of being told this, he tries to dissuade dealers from calling. The dealer placing the advertisement may sell the clock from his own home address, or more usually from a different address, perhaps that of a relative whose surname is different from his own. As often as not by advertising the item as a private sale, the vendor can avoid any redress should the buyer later find the item faulty. Also of course he can probably avoid paying tax on the sale. Naturally, there must be perfectly genuine private sales from time to time, but if you are thinking of buying a clock in this way, be very much on your guard.

Not long ago I was talking to a young couple who were very enthusiastically telling me about their newly purchased clock. They had been 'put on to' this by the friend of a friend of a friend. They had bought the clock from an eccentric old gentleman 'collector', who did not really buy and sell clocks, they assured me, but who had a large collection. Apparently, if this old gentleman took a liking to you, he might be willing to sell you one of his clocks. He must have taken a great liking to this young couple for not only was he prepared to let them have one of his clocks, but was also willing to take in part exchange an old wall clock which they had and no longer wanted.

They asked me if it was possible to get the moon and date dials to function on their newly acquired clock, as they did not seem to be working. I arranged to call and see the clock, and when I did I found a very nice case, a very pleasant brass dial, and a nice enough movement. The only problem was that none of the three was made for either of the other two. The dial had been made for a one-handed clock, but it now had a two-handed movement which fitted so badly that the teeth had had to be cut away from about a quarter of the circumference of the date and moon discs to prevent the discs from obstructing the hands pipe. I explained this to them and showed them where the teeth had been cut off. They then asked me to value the clock. 'Which piece?' I asked. 'The dial, the movement, or the case?'

These people were a fairly prosperous, educated, and apparently sensible young business couple who could well have afforded to buy a clock from any local dealer. But instead of paying the going price, they had had to make a 200-mile round trip with the clock on the roof of their car, had been sold a faked clock as well as having lost what might have been a quite reasonable one taken by the 'collector' in part exchange, and heaven knows how much money they had been persuaded to part with at the same time. This story is absolutely true, and the incident took place only recently. One would imagine that anyone would have been slightly suspicious if two or three dials of a clock were not working, and when the simplest glance behind the dial revealed that half the gearing had been chopped away.

It is worth remembering that any dealer has some sort of reputation to maintain. Obviously, some are better than others, but with a dealer you always have some sort of redress. This applies especially if you insist on some sort of guarantee about a clock you are buying, and most of the better dealers will give this anyway, without having to be asked.

There is another side to the coin, however. There are very many dealers and specialists who, far from cheating anyone, are genuinely enthusiastic about the goods they handle. They are usually pleased when a customer asks intelligent questions or expresses sensible opinions. There are customers who walk into a shop, ask which is the best clock, and promptly buy it with no further conversation. That sort of sale falls rather flat for an interested dealer, who likes to think that the customer will appreciate the item he is buying, and will ask questions about the clock, its age, its style and maker, as well as just signing a cheque.

It does not generally take long to appreciate which dealers are sincere and which are not. A good dealer is not some sort of high-powered salesman. He does not try to impress you with talk about how many clocks he sells in a month; how this one is a bargain; how the Victoria and Albert Museum are considering buying that one, so you had better be quick if you want it; how this one has just come into stock, and no one else has had a chance to see it yet; or any other gimmicky sales patter calculated to rush you, fluster you, embarrass you or cajole you into buying something that you don't want and which he wants to be rid of. You will often find that the most garrulous and ebullient salesmen are the ones who dry up instantly when you ask an intelligent question, and who will then hurriedly explain that they don't really understand clocks very well, but just sell the occasional one now and then.

Buying a clock is something you want to do pleasurably, unhurriedly, calmly, quietly, considering carefully the height, colour, style, period and price in relation to your pocket, your home and the setting for the clock. The final decision should always be a contented one. If you are not absolutely happy in your own mind about the prospective purchase, then don't buy it. When you see the clock you want to live with, you will know it.

I remember one occasion when I sold a clock to a customer who lived not very far from me. He was delighted with his purchase, so much so that later on he called in a couple of times to tell me, and eventually he brought round a friend and his wife to show them my clocks, as they had admired the one he had bought. He thoroughly enjoyed himself showing them the items, and they in turn were captivated. I hardly had a chance to get a word in! They chose a clock which suited them, I delivered it, they paid me: all was settled and done. A week went by. The telephone rang. Could I possibly agree to take back the clock? No, there was nothing wrong with it; it was very nice and going perfectly. It was just that his wife did not like it in that particular place, and he did not like it where she liked it; and he had had to move his stereo record cabinets, and then couldn't get to the electrical socket, and so on and so on. I collected the clock and gave him back his cheque. The trouble was that they had not given proper thought to their purchase. They had liked the clock and the idea of having it, but they had not considered how it would fit into their home; and the real answer was that it did not fit in. They had allowed themselves to be carried away on the tide of their friend's enthusiasm, instead of considering carefully and calmly. Don't make the same mistake.

There is one thing which you as a clock owner, potential purchaser or admirer owe to yourself, and that is to hold fast to your own tastes. By all means learn the 'whys' and 'wherefores' of clocks, but let no one tell you what you *ought* to like. A clock, more than any other piece of furniture, has a character, a personality, of its own. You have to live with it in your home, and a clock which you find attractive will give you many years of pleasurable company. Each day you will find yourself giving it a little nod of approval and appreciation from the sheer pleasure of looking at it. Therefore pay no attention at all to what someone else happens to like, whether or not he is an 'expert'. Do not let others hinder your judgement with their prejudices. Try to look with an open mind and you will recognise what you like as soon as you see it. A book may teach you how to look at a clock, what features to notice, and whether these features are well executed. It may try to advise you about 'quality', a much-used word in the world of antiques. So often this word is used to signify something which was difficult to make and costly in labour and materials, and therefore supposedly good. But however much of this kind of 'quality' a clock may have, this may by no means make it beautiful. All the 'quality' in the world cannot compensate for ill-conceived design, and of course the design is relevant only to the type of clock in question. It is quite unfair to compare a cottage clock with a palace clock as either would fail in the wrong setting. Similarly, do not be misguided in your attitude to white dial clocks by glowing descriptions of the slender and elegant case styles of the seventeenth-century London clockmakers with names like Tompion and Fromanteel. Beautiful as their case styles may have been, these clocks belong to an entirely different period and setting; and no white dial clock was ever intended to look like a seventeenth-century London 'coffin' clock. By all means learn to like clocks, but do not let others force their tastes on to you.

CHAPTER 14

RESTORATION, RESUSCITATION AND REJUVENATION

The basic principle behind the restoration of any antique object is to try to put it back into its original state, as far as possible by restoring the existing (if original) parts, replacing only those parts which are beyond restoration, and making any essential replacements in the correct style. If 'restoration' oversteps these limits it can soon become 'faking'. For instance, the converting of thirty-hour clocks to eight-day ones is not restoration, it is faking. It is furthermore a great pity, but it is widespread for all that. An essential feature of an antique, and one especially applicable to clocks, is the complete absence of a coating of dirt, grease, cobwebs and general grime: all of these, apart from spoiling the appearance of the clock, will grind themselves into the moving parts and help to shorten its life considerably. A clean clock is likely to work better and to last longer.

There are good clock restorers and bad ones. One would think that it would be quite unnecessary to point this out, but I am frequently astonished by the appallingly low standard of restoration which some people accept. As long as a clock ticks, that seems to be adequate for some. Often in conversation I hear about a clock which has just been 'restored' or repaired, but if I ask specially, I often learn that the date indicator does not work, the moon dial does not work, and sometimes even the strike does not work. I personally have seen very bad examples of this even from people who describe themselves as 'horologists' and who belong to trade associations.

Date dials and moon dials can be tricky things to put right, and if they do not function precisely they can cause the clock to stop. Hence some repairers do not wish to fiddle with dates. They may perhaps tell the customer that the date indicator cannot be made to work, and leave it at that.

The answer of course is that the customer should refuse to accept this. If one repairer cannot undertake to make the clock operate fully, the customer should take it to a better repairer who can. You would never dream of patronising a garage where they could repair engines but could not repair faults on, say, brakes or lights or windscreen wipers. One should be quite clear about this: if a clock was made with a particular feature, such as a moon dial, this was made originally to work and should be capable of being made to work today. Therefore do not be prepared to accept that a certain feature is irreparable without your being given a very convincing explanation. There is no such thing as a clock which cannot be repaired, though there are occasions when the cost of the repair work could be prohibitive.

Recently I saw a very valuable clock, the dial of which had an indicator for the (seven) days of the week and the (twelve) months of the year. Behind these sub-dials belonged wheels with seven and twelve points respectively. Incredible as it may seem, the original star-wheels had been put back behind the wrong dials by a 'restorer' who, having experienced difficulty in getting the seven-pointed wheel to fit behind the twelve-month sub-dial, had set about filing part of this away to make it fit!

Many good restorers, and there are plenty of them, will not take on repair jobs on dirty clocks unless they can also clean the movements. This is understandable, since a movement that is clogged up with grime and oil can hardly be expected to function well for long, and if it should stop by being clogged up with grime, the customer might blame the workmanship of the restorer. You cannot have it all ways. A repairer who is good will tend to be more expensive than one who is inefficient, but you get the workmanship you pay for, and it is probably cheaper in the end to pay a capable man once than an incompetent one several times.

If a repairer is a member of a professional association, you may feel that you could have more confidence in him. At least, if such an association member fails to give satisfaction, you have the opportunity of taking the matter up with the association concerned.

One word of caution: do not fiddle about with a clock. Left to itself a clock will often function happily for many years and, surprisingly enough, even sometimes when it is in a very dirty condition. Clocks do not often wear out under normal conditions. If a clock does not go, it is very often because someone has broken it. I have been told many times by people who have newly moved into a house that their clock used to go in the old house but will not go in the present one. Often it is simply not level, or else the fork holding the pendulum suspension spring has been bent during removal. When one hears of removal men tipping a clock on to its side and carrying it out lengthwise, it is hardly surprising that it will not go in its new home. Carrying a clock in this way without dismantling it is almost certain to break it. You must dismantle it by removing first the hood, then the pendulum and weight(s), and then lifting the movement clear of the case. It is a very simple operation, but if in doubt it would be as well to get a clock restorer to dismantle it for you—rather than have to pay him to mend it later!

Do not wind clock hands backwards, and do not wind them forward without allowing each hour time to strike fully as you go. Do not let your clock run right down. If, for instance, you are going away on holiday, stop the clock before you go. If you allow it to run down fully the sequence of striking may be put out and you may then have to reset it yourself: far simpler not to let it get out of sequence in the first place. All these things may seem obvious, but sometimes it is the obvious things which are overlooked.

White dial clocks sometimes suffer damage to the dial as a result of rusting from beneath the paint surface, and ultimately pieces of the surface may chip off like eggshell or even peel off. For those whose clocks have this problem there are specialist firms of dial restorers who can do remarkably good cleaning and restoration jobs. If you are thinking of buying a clock which has this fault, then bear in mind that such restoration is going to add to the bill. However, if the clock is a family heirloom, it can be saved and made to look just as good as the day it was first sold. In the same way, a dial on which some of the lettering or numbering has worn faint can be restored to its former clarity.

Restoration of casework is exactly the same as for any other piece of antique furniture. An interesting aspect of case restoration is that of pine cases, or deal cases if you prefer the term. It was called deal in the past but today most people refer to deal furniture as pine, stripped pine or waxed pine—it is all the same thing really. A deal case was often designed for a less distinguished position than one in, say, oak or mahogany. It was for a clock case for a kitchen perhaps, or a farm cottage, or the servants' quarters in a larger house. It was the cheapest wood and was used for economy. As such, a pine case generally lacks brass fittings in the form of pillar caps and paterae, etc, and where possible, turned wooden alternatives were used, though not always.

This is not to say that a pine case is now regarded as an inferior item. On the contrary, it often has a particular charm because of its simplicity; and many pine cases are very attractively proportioned. In fact, in the absence of elaborate detail, proportion becomes a very much more important factor than in, say, a case where the beauty lies mainly in the veneers.

The interesting thing about pine cases, however, is that when they were made they were usually sold *painted*, or sometimes stained. I say 'usually' though I am tempted to say 'always', as I have never seen a pine case that does not show some signs of having been painted or stained, and very often some traces of red priming paint remain. Furthermore, the wood used is often knotty and the nailheads are crudely hidden with filler, making it apparent that the bare pine surface was never intended to be seen. Now stripped pine is very popular at the present time, both in antique furniture and modern furniture, and especially for kitchens. For this reason pine-cased clocks are popular, but they are not too easy to come by in good condition as pine cases tend to be prone to woodworm attack. Most dealers today who buy a painted, pine-cased clock would strip off the old blackened paint and sell the clock as 'stripped pine'. If you ever saw a clock of this type with shrivelled, blackened paint still on it, you would hardly take to it in that state.

If we were being strictly correct, what we dealers ought to do is to repaint pine cases. But what an outcry there would be if a dealer did this! We—customers and dealers—agree to the fair and sensible restoration of clockcases, except for pine ones, where we all agree that they should be stripped down to the bare wood and waxed, thereby producing an effect totally different from the one intended by the clockmaker who first sold the clock. It really is rather an odd situation. Ironically, it is often only because it was painted, that a pine case has survived being devoured by woodworm. I do not mean to indicate that I disagree with stripping pine cases. I am simply pointing out that in treating them in this way we are breaking the rules of restoration. The cost of stripping is probably twenty times that of repainting. And as a final irony, in twenty-five years' time antique dealers will no doubt get out the paintpots and repaint all the pine cases!

'Patina' is a word much bandied about in antique circles, especially if one is trying to sound knowledgeable. A clock case which has been sadly neglected may have lost not only its 'patina', but perhaps its colour too. You can obtain all manner of weird and wonderful waxes, and polish until you are 102, and you may be no nearer to restoring either the colour or the surface texture to a desirable state. An oak case may have been kept in a damp outbuilding: its surface has been attacked by moisture, which has broken through the wax coating and into the varnish beneath, and has attacked the original stain, causing the oak to appear now as a greyish white colour, maybe in patches. Where traces of varnish remain, these may be blistered into tiny bubbles and may well have turned white. You will be extremely lucky to manage to camouflage bad damp-staining by rubbing in waxes. The energetic application of brown boot polish is likely to be more profitable to boot-polish manufacturers than beneficial to clock cases.

I am emphasising two points. First, the oak case originally had colour applied to it when new. It was not, with very few exceptions, sold as plain white or yellow oak; it was stained initially to give it some richness and colour. Often it has a rich red tint to it. This colour may have improved through years of care, but the colour was put into it to begin with.

Second, after staining, the case was *varnished* prior to being sold. This was to keep in the colour and to provide a sealed base on to which one could wax. If you try staining without varnishing, the colour will gradually come out or be considerably weakened when wax is rubbed in. With later veneered cases the finish is often akin to a thin French polishing, perhaps a shellac coating of some kind. Scratch the surface with your fingernail and you will see this hard layer shelling off.

Grandfather clocks were not revered in the past as they may now be: 99 per cent of them stood in ordinary households and received every bit as much wear and tear as any other piece of furniture—and that, in most houses, meant vastly rougher treatment than most of us would give our possessions today. Our grandparents could renovate their old furniture, if they could afford it, by French polishing. For those of smaller means the varnish pot was the answer, and it seems that varnish was often used to 'brighten up the old clock a bit'. Both these forms of treatment are today considered to have been misdirected enthusiasm.

I am not advocating the rubbing down and repolishing of cases needlessly. However, if a part or even all of a clock case is in very bad surface condition, or treacled with several layers of varnish, it may be necessary to remove the old, shrivelled varnish, restain, re-varnish lightly and then rub down with fine wire wool prior to waxing. This at least gives one a chance of working up some sort of surface texture which will improve with the years as the wax hardens.

CHAPTER 15

HOROLOGICAL REFERENCE BOOKS

Shortly after the publication of my book on Yorkshire clockmakers, I recall chatting one day to an antique dealer about the book. 'Nay, lad,' he said, 'I've been buying and selling clocks all these years—what would I want with a book like that?' A down-to-earth question, for which I had no answer. As far as his interest went, of course, he was quite right. He knew how much to pay for particular types of clock and how much to sell them for. As to anything else, he did not really care who had made the clock or when. He knew from his own experience what no book could tell him: prices and saleability. Fortunately the majority of antique dealers have a genuine interest in the objects they handle, additional to, and in some cases more important than, the daily business of buying and selling.

In pursuit of your interest in clocks, whether in buying, selling, or just looking, there may be times when you will wish to turn to some of the basic horological works of reference, and a few suggestions might prove helpful.

As basic background material for longcase clocks you ought to read E. L. Edwardes's *The Grandfather Clock*, 1972 edition. Another book for general background is Eric Bruton's *The Longcase Clock*. Both these books give considerable space to country longcase clocks, which one might normally meet with, as well as discussing top London-made clocks, which are beyond the financial aspirations of most of us.

When it comes to reference books giving lists of makers to help in dating a particular clock, there are several works written specifically for this purpose. By far the most helpful and detailed ought to be those books which are purely about the makers in individual counties, and several counties have been covered in this way. Some of these county works are now out of print, but should be available through a library. Books I know of this type are:

Clockmaking in Oxfordshire by C. F. C. Beeson, Oxford, 1967.
Colchester Clockmakers by Bernard Mason, London, 1969.
Cornish Clocks and Clockmakers by H. Miles Brown, Newton Abbot, 1970.
Devonshire Clockmakers by J. K. Bellchambers, Torquay, 1962.
Somerset Clockmakers by J. K. Bellchambers, London, 1968.
Yorkshire Clockmakers by Brian Loomes, Clapham (via Lancaster), 1972.

Research into the clockmakers of a number of English counties is currently in progress, and the following works are known to have reached an advanced state of preparation and will probably appear in print before long.

Alan Treherne is working on books on the clockmakers of Cheshire, Shropshire and Staffordshire. A. W. Branston is working on Herefordshire. *Suffolk Clocks and Clockmakers* by Arthur L. Haggar and Leonard F. Miller is expected to appear in print shortly. John B. Penfold has virtually finished his book, *The Clockmakers of Cumberland.* I myself am well under way with books on the makers of Lancashire and Westmorland and have started work on Northumberland and Durham makers. Clive Osborne is working on Essex clockmakers and has gathered much data on makers from many counties. These next few years should produce some of the most interesting books on localised clockmaking, and generally these books on particular areas are where the real information lies.

In the meantime, for those areas not yet covered individually, one has to turn to one or more of the following: *Clock and Watchmakers in Wales* by Iorwerth C. Peate; *Old Scottish Clockmakers* by John Smith.

Baillie's *Watchmakers and Clockmakers of the World*, and the list of clockmakers in Britten's *Old Clocks and Watches and Their Makers* give poor coverage of provincial makers.

There are some important factors that should be fully understood when using any horological reference books, and especially when using those books of a general nature. With the individual county lists the authors have frequently been able to ascertain the dates of birth and death of the makers listed, though not, of course, of every maker in such a book. Often, however, the lists may give a single date for a maker, simply a known year when he was working. A single date shown thus can be almost as misleading as helpful, since a clock by that man could have been made as much as fifty years distant from the given date. It is important to understand that unless positive dates of birth and death are stated, any date given should be regarded as only a rough guide to the period when the man is known to have been working.

Often when a single date is given, it is preceded by the word circa or the abbreviation c. This of course means 'approximately' or 'about', and is usually an indication that the date is simply a guess or an opinion as to the period of a clock. How valid that date is will depend very much on the competence of the person whose opinion it represents, and that could be a very variable factor.

If you were doing a crossword puzzle, you might later check your answers against the correct solutions. Similarly, after attempting to date a clock, you might then think of checking with a book such as Baillie to see if you were right. Where such an approach falls down is that the reference book you are checking by is simply another person's attempt at the crossword, not the definite answer. Learn to date clocks for yourself from your own judgement. If some reference list of makers confirms your dating, well and good. If it does not, this need not mean that you are wrong. One does not need to have been looking at clocks for long to realise that by no means all the answers are in books, and by no means all the answers that are in the books are the right ones.

Similarly, you should not feel that because a clockmaker's name appears in a reference book, this in some way means that he is a better maker than one whose name fails to appear in that book. It is in fact simply an indication that the writer may have come across work by the one man and not the other. The absence of a maker's name indicates, if anything, a failing on the part of the author rather than the clockmaker. Yet people *do* tend to have more faith in a clock whose maker appears in the standard lists. There is somehow a feeling that a clock by a known maker may be more genuine, or more authentic, or more safe, while an unknown name does not inspire confidence. In fact when you think about it, the reverse applies. There would be absolutely no point in *faking* a clock by an unknown maker, and if the buyer is to have any doubts at all, he would be far wiser to have them about a clock bearing a well-known name.

This concern with known names reminds me of a little incident of some interest. Quite recently I bought an eight-day white dial clock of about 1790 or so. The dial was very worn and the maker's name illegible. After about an hour of very careful examination under a strong light I was able to decipher the name, though only with great difficulty. It was important to me to know the name so that I could have it relettered along with the numbering when the clock was restored. Finally I made it out: Forster of Sheerness (Kent). I was very pleased to have been able to recognise the maker, and promptly, as I suppose many of us do on such an occasion, I looked him up in Baillie. The entry read: 'Forster, . . . Sheerness, ca. 1790 CC'.

This means: Christian name unknown, working about 1790, member of the Clockmakers' Company of London. Having found this I was even more pleased, but there was really no reason for me to be. The magic letters *CC* had done it. The assumption (and of course it is an unfair one to make) is that because a man was a member of the Company he was somehow a better clockmaker than one who did not belong to this body of (largely London) clockmakers. It was quite wrong of me to attach any more importance or value to this clock because of that fact.

The clock had a Birmingham-made dial and falseplate. It was no better or worse than a thousand others just like it. If it had been Forster of Plymouth or Manchester or Dundee it might have looked identical. And here is an even more interesting thought. If some unscrupulous dealer had arranged for that clock to be relettered, and had deliberately decided that it was to be by 'Forster of London', there is no one who would have known except that dealer himself. There was nothing about that clock, or thousands of similar ones, to say who had made it or sold it, except that name on the dial.

The real point I am making is that since an average white dial clock is almost identical, regardless of where it was made, with others of the same period and style, there is no good reason why a London-made one should be valued more highly than one from any other town.

There are times, I feel, when a person buying a clock wrongly hankers after 'quality'. He feels it would be better to have a clock by a maker who is said in some book or other to be a 'good maker'. So he reads a book or two, and perhaps sees, for example, that Thomas Lister of Halifax, Yorkshire (there were two, by the way—father and son), has a good name and that he made some clocks of very ingenious and fine workmanship. Having read this he sees a clock bearing the name of Lister and buys it, assuming that he has bought a *good-quality* clock. In actual fact, while it is true that this maker (the son) produced some very fine clocks, it is also true that his average clock was no better and no worse than anyone else's of that time. This is because a clockmaker made what his customers wanted; and the fact that a man made some good clocks by no means indicates that he did not make many very ordinary ones too. Look for a 'quality' clock if you wish, but do not assume that the quality exists because of the name on the dial. I am speaking here only about clocks by provincial makers. I am not competent to say whether or not the same applies to clocks by early London master-clockmakers. As we have just seen, the name on many white dial clocks is hardly important anyway, but in this paragraph I am thinking of brass dial clocks as well. If it is a 'quality' clock you seek, by all means buy one, but make it one which you yourself consider to fit that description, not one that someone else tells you is good. Remember that no clock dealer will boast of selling clocks of *poor* quality!

One final point about checking on the dates of a maker. People often tell me that their white dial clock has no name on it. Occasionally they may be right but in my experience probably less than 1 per cent of the clocks are nameless. There may at first sight appear to be no name on the clock, but on closer inspection one can usually see that the name has been worn off by over-cleaning through the years. The way to decipher a very faded name is to get a torch or bright light of some sort and hold it close to the dial at an acute angle. Very often one can then make out the name of the maker and the town. White dial clocks almost always have the town as well as the man's name, if you have the patience to read it. There are occasions when dials have been *restored* very crudely and the names touched in incorrectly. In these instances skill and patience are needed in the deciphering. Finding out about the age and the maker of a clock is a fascinating pursuit for its own sake and half the fun of owning a clock lies in knowing about its history.

Still on the subject of worn names on dials, I was horrified to read recently in a new book (*English House Clocks* by A. Bird) the statement that if the original name on the dial could not be made out, 'it may be proper' to write on the name of some 'unimportant' maker. The old-fashioned and incorrect implication in this is that 'important' makers did not make clocks with painted dials. Alternatively it is suggested that one might paint on a fictitious name and address. Thank heavens not many writers openly advocate a practice of this kind. If they did we might soon find that it was recommended standard practice for picture dealers to paint some 'unimportant' name on unsigned canvases or for antique furniture dealers to mark their pieces with some 'unimportant' name. Before many years had passed a whole literature would have been built up of names from such pieces, and actual examples would be on record as having been made by craftsmen who never existed. Enough problems exist for those of us who are interested in research into past craftsmen without these being multiplied by such scandalous forms of so-called 'restoration'.

SELECT BIBLIOGRAPHY

From the extensive published literature on horology I feel that the following works may be of interest to those wishing to read further into the subject:

Baillie, G. H. *Watchmakers and Clockmakers of the World*, 1969
Beeson, C. F. C. *Clockmaking in Oxfordshire*, Ramsgate, 1967
Bellchambers, J. K. *Devonshire Clockmakers*, Torquay, 1962
——. *Somerset Clockmakers*, 1969
Brown, H. Miles. *Cornish Clocks and Clockmakers*, Newton Abbot, 1970
Bruton, Eric. *The Longcase Clock*, 1970
Cescinsky, H., and Webster, M. R. *English Domestic Clocks*, 1913 (repub 1969)
Edwardes, E. L. *The Grandfather Clock*, Altrincham, 1971
Goodison, N. *Gillow's Clock Cases*, 1968
Loomes, Brian. *Yorkshire Clockmakers*, Clapham (via Lancaster), 1972
Mason, Bernard. *Clock and Watchmaking in Colchester*, 1969
Peate, Iorwerth C. *Clock and Watch Makers in Wales*, Cardiff, 1960

Mr Kenneth Roberts has kindly supplied the following list of books, which might prove helpful in identifying American clockmakers of this period:

Chandlee, Edward E. *Six Quaker Clockmakers*, Philadelphia, 1943
Drost, William E. *Clocks and Watches of New Jersey*, Elizabeth (NJ), 1966

Eckhardt, George H. *Pennsylvania Clocks and Clockmakers*, New York, 1955

Hoopes, Penrose R. *Connecticut Clockmakers of the Eighteenth Century*, Hartford (Conn), 1930

James, Arthur E. *Chester County Clocks and Their Makers*, West Chester (Pa), 1947

Palmer, Brooks. *The Book of American Clocks*, New York, 1950

——. *A Treasury of American Clocks*, New York, 1967

ACKNOWLEDGEMENTS

The number of dealers or even enthusiasts who can date a white dial clock accurately from the dial style alone is not large at the time I write this. Not because it is difficult to do, but simply because few have taken the trouble to look closely enough at the details. I have received little encouragement from some of the people I had hoped might be most interested, and considerable support from some areas where I least expected it. I must acknowledge the interest shown by many clock owners who have written to me during the preparation of this work. Mr W. A. Taylor, Birmingham City Librarian, went out of his way to be helpful. My colleague, Mr Peter C. Nutt, assisted with some of the genealogical research, as he has done on many occasions for me.

I must especially thank Mr E. L. Edwardes, author of *The Grandfather Clock*, for kindly reading through the manuscript and for his considerable interest and constructive suggestions. I must also acknowledge the assistance of Mr Kenneth Roberts, formerly of the American Clock and Watch Museum Inc, Bristol, Connecticut, for the same. Both these gentlemen are deeply and enthusiastically involved in horological researches of their own, and because of that I am all the more grateful to them for sparing time to help with mine.

Thanks are due also to my friends Brian Morison and David Barker for their enthusiastic interest and help; to Messrs Manby of Skipton, ironmongers, for allowing me to examine their old business records; to Miss S. J. MacPherson, BA, Archivist in Charge at

Westmorland Record Office, Kendal, for assistance with the records of stock transfer of Jonas Barber III; to Mr A. Ankerson, Mr Eric Benton, Mr John Coker, Mr Jeremy Evans, Mr Bernard Mason, OBE, and Dr J. M. Plowman; and to many others too numerous to mention by name.

Most of the illustrations were taken by the author and are of clocks which have passed through his hands. However, author and publishers wish to thank the following for permission to publish plates of clocks in their possession: 13 and 31, Mr and Mrs K. Gooch; 15 and 34, David Barker; 5 and 36, C. Williams; 18 and 41, Mr P. Aram; 6 and 26, K. V. Flanagan.

Plates 18 and 41 were taken by Malcoms Photo Services, Otley.

INDEX

The summary tables of dial and case features on pp 96–7 and 136–7 refer the reader to the relevant period and thus to the appropriate part of the book, and these stylistic features are therefore not included in the index.